The Robot Era

P. E. Cleator

Routledge
Taylor & Francis Group
ROUTLEDGE REVIVALS

First published in 1955
by George Allen & Unwin Ltd.

This edition first published in 2025 by Routledge
4 Park Square, Milton Park, Abingdon, Oxon, OX14 4RN

and by Routledge
605 Third Avenue, New York, NY 10158

Routledge is an imprint of the Taylor & Francis Group, an informa business

Publisher's Note
The publisher has gone to great lengths to ensure the quality of this reprint but points out that some imperfections in the original copies may be apparent.

Disclaimer
The publisher has made every effort to trace copyright holders and welcomes correspondence from those they have been unable to contact.

A Library of Congress record exists under LCCN: 55004837

ISBN: 978-1-041-16444-9 (hbk)
ISBN: 978-1-003-68462-6 (ebk)
ISBN: 978-1-041-16446-3 (pbk)

Book DOI 10.4324/9781003684626

FIRST PUBLISHED IN 1955

PRINTED IN GREAT BRITAIN
in 11 point Times Roman type
BY ROBERT MACLEHOSE AND CO. LTD
THE UNIVERSITY PRESS, GLASGOW

To Robot No. 1600
(MDC)

★

Foreword

THE essence of the Industrial Revolution was a replacement of animal muscles by machines. This mechanisation is now being followed by a process of automatisation, a development which promises to relieve mankind of much mental effort, particularly of a routine nature. Not only do machines exist which can take-over the steering of ocean liners and direct road traffic, but there are other mechanisms which can read blue prints, or answer a telephone call in a person's absence (and take a message), or react intelligently to the spoken word, or make lightning translations from one language to another.

As yet, of course, all such cogitative contrivances occupy a very lowly position in the domain of thought and meditation. No machine has yet been designed, much less constructed, which even approaches in complexity the bewildering intricacy of the human brain—though this is not to say that none ever will. In the meantime, it is a fact not without significance that in certain respects, *e.g.*, speed of operation, the mind of man has been surpassed by some of the mechanical substitutes of his own devising. Electronic computers can carry out in a matter of minutes highly complex calculations which a competent mathematician would take months to complete—the British Ministry of Supply has recently disclosed the existence of an analogue machine, housed at the Royal Aircraft Establishment, Farnborough, the computational exertions of which are equivalent to the tracking and plotting activities of no less than 10,000 Service personnel!

Such machines can not only check the result of their operations on completion (or as the work proceeds), but, aided by an infallible memory system, they can also learn from experience, adding progressively to their store of knowledge. In the days ahead, *Homo sapiens* will have increasingly to contend with mechanisms so mnemonically endowed that they can never forget! Such machines now exist. Indeed, they are already in use. And they may well be the forerunners of a mechanical intelligentsia which in time to come will challenge the mental supremacy of man himself. . . .

P. E. C.

HESWALL, 1955

Author's Note

THANKS are due to all those members of the many concerns and official establishments who assisted in the preparation of this writing by their ready response to requests for information, literature, and photographs: a comprehensive list will be found on pages 158–160. In addition, I have also to acknowledge the help given by the Librarians of the University of Liverpool and the Agricultural and Mechanical College of Texas; by A. W. Kinnard, Assistant to the Master of the Armouries, H.M. Tower of London; by Newall Rogers, New York Manager of the London *Daily Express*; by Peter J. B. Stevens, Foreign Editor of *Design News*, Detroit; by J. L. Nolan, of the Information Service of the U.S. Embassy, London; and by John L. Atkins, of the Foreign Service of the United States of America, London.

A number of firms were good enough to permit a visit to their premises, and it is with pleasure that I place on record my appreciation of this privilege, and acknowledge the help given by E. Heys and his colleagues, of the Metropolitan-Vickers Electric Company, Limited; by H. Busby, of Cincinnati Milling Machines, Limited; by F. V. Ault, of the Ballito Hosiery Mills, Limited; by F. R. Ashworth, of the British Manufacture and Research Company, Limited; by H. M. Robinson, of Ind Coope and Allsopp, Limited; by J. Bowman, of the Austin Motor Company, Limited; by I. C. McDougal and S. Smith, of Tate and Lyle, Limited; by R. L. Michaelson, A. G. Coaten, and E. M. Webb, of the British Tabulating Machine Company, Limited; by I. Wright, of Glass Bulbs, Limited; and by A. L. Paget, of the General Electric Company, Limited.

I am further indebted to the six last-named of the above concerns for checking my descriptions, a service in the interests of accuracy which was also rendered by The Automatic Telephone and Electric Company, Limited; by Mavor and Coulson, Limited; by the Esso Petroleum Company, Limited; by British Hartford-Fairmont Limited; by Electrolux Limited; by Cinema-Television, Limited; by Tyer and Company, Limited; by General Motors, Limited; and by the B.B.C.

A small proportion only of the many photographs received could be used, and the source of those selected is given in each case. Regarding other illustrations, two of the line drawings—Figs. 6 and 7, depicting the Metrovick hydraulic control system and the Electrolux cycle, respectively—are based on diagrams contained in literature placed at my disposal, and the adding circuit shown in Fig. 10 is taken from R. L. Michaelson's Paper, *Large-Scale Electrical Digital Computing Machines* with the kind assent of the Author, and of F. H. Wales, Joint Editor of *The Journal of the Institute of Actuaries*.

For permission to reproduce the *Sunday Dispatch* cartoons, I have to thank that publication, its Picture Editor F. T. Pigrome, and the various artists concerned—F. Holmes, Bruce Cavalier, J. F. Kirkbride, A. L. Teal, and W. Beckett (Maskee). The two sketches by Clew and Wimsey, from the *600 Magazine*, and the David Langdon and Crawford drawings, from *Lilliput* and the *Daily Sketch* respectively, appear through the courtesy of the Editors concerned. The illustration by Trez was supplied by the Artist, the Brockbank sketch is reproduced by permission of *The Motor*, and Sabo's *Cavalier* cartoon appears by arrangement with the Ben Roth Agency, New York.

As usual, my wife has been called upon to assist in the task of proof reading, and once again I find myself indebted to C. A. Furth—not only for inviting me to write the book, but for the deletions and additions which he counselled in the interest of its improvement.

Contents

Illustrations

PLATES

FIGURES

CARTOONS

PART ONE

Wheels Within Wheels

★

I

Matters Mechanical and Metallurgical

WE all believe we know what a machine is until we are asked to define it, whereupon assurance is apt to make way for doubt and uncertainty, if not actual dismay. Precisely what is it that is common to a bicycle and a steam engine, an electronic computer and a lawn mower, an airplane and a village pump? The answer is short and to the point: a capacity for doing work. A machine, that is to say, is a device for transferring or modifying energy, and as such it is represented no less by a garden spade than by a Diesel locomotive or a rotary printing press.

Six so-called simple machines have long been recognised, but of these, the pulley and the wheel-and-axle may be considered modifications of the lever, and the screw and the wedge can be regarded as variations of the inclined plane. In effect, the number of basic machine forms reduces to two—the lever (characterised by an equilibrium of torques) and the inclined plane (which involves a vector resolution of forces).

The lever, as exemplified by the crowbar, is the most famous of all machines. It has been known, in one guise or another, since time immemorial, and the probability is that it is almost as old as man himself. Some primitive forebear, no doubt, first happened upon it while amusing himself with a stout branch purloined from a fallen tree. We can picture him playfully inserting one end of his improvised staff into a gap between two adjacent boulders. We can imagine his astonishment and delight at finding that his lifting powers were thereby increased no less than tenfold. And we can anticipate the solemn warning of his union leader, that such inexplicable levitation threatened the livelihood of all the members of the confederation of master weight lifters and quarry-men.

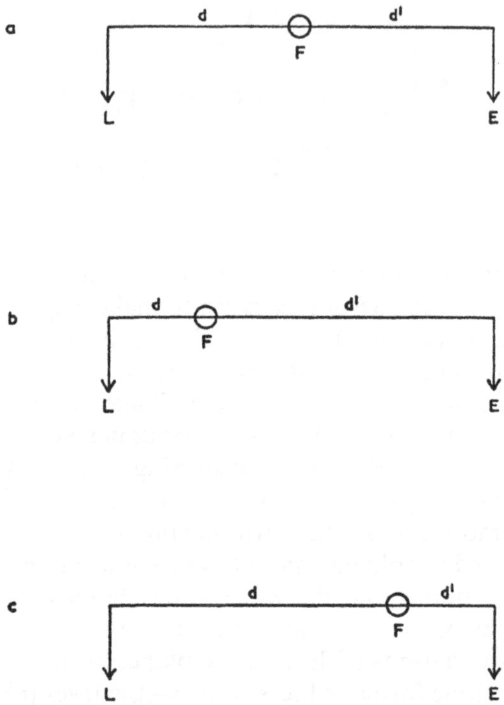

FIG. 1.—Class I levers—load and effort on opposite sides of the fulcrum. There are three possibilities: (*a*) $d=d'$; (*b*) $d<d'$ (read, '*d* is less than *d'*'); and (*c*) $d>d'$ (read, '*d* is greater than *d'*'). Depending upon whether $d=d'$ (as in the ordinary balance), or $d<d'$ (as in a pair of pliers), or $d>d'$ (as in the steelyard), the mechanical advantage is respectively 1, greater than 1, and less than 1.

The lever is essentially a rigid rod which turns on a support, *F* (the fulcrum), whereby a load, *L*, acting at one point of the rod, can be overcome by an effort, *E*, applied at some other point*

* It will be clear that when the length of the lever's two arms (designated *d* and *d'*—Figs. 1 and 2) is equal, there will be equilibrium if the load and the effort are also equal, and that, in general,

$$\frac{E}{L}=\frac{d}{d'} \text{ or } L\times d=E\times d'.$$

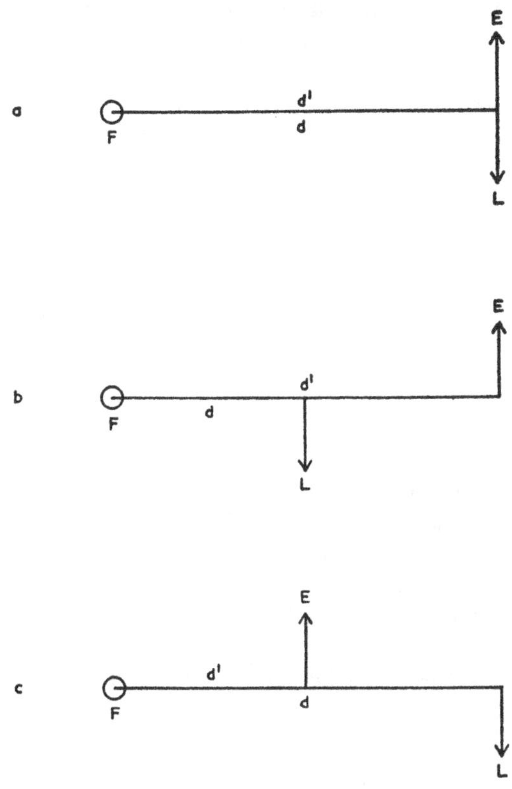

FIG. 2.—Class II levers—load and effort on the same side of the fulcrum. A familiar example of (*b*) is provided by the wheelbarrow in which $d<d'$, and the mechanical advantage of which is thus greater than 1. In (*c*), the so-called Class III lever (typified by the human arm), the opposite is the case. Here, $d>d'$, and the mechanical advantage is accordingly less than 1.

(Figs. 1 and 2). What is termed the mechanical advantage of the device is readily determined by dividing the load by the effort. It follows that when the load and the effort are equal, the mechanical advantage is 1, that when the load is twice the effort, the mechanical advantage is 2, and so on.

As for the inclined plane, this was originally discovered, perhaps, by some mover of megaliths whose workmen contrived to raise the site level of an outsize specimen by the revolutionary expedient of dragging it up a slope* (Fig. 3). Be it noted, how-

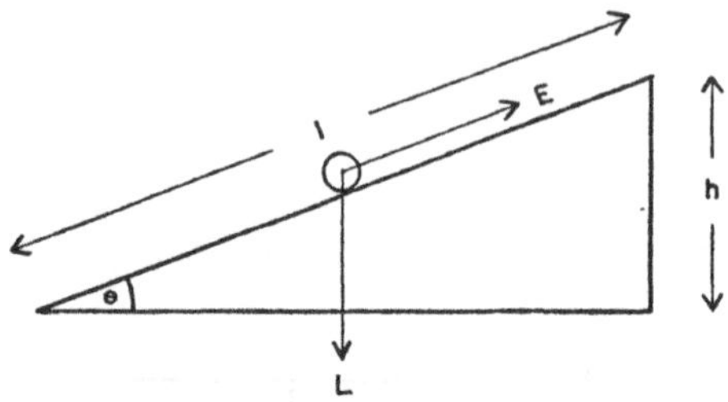

FIG. 3.—The mechanics of the inclined plane. With a load of 144 pounds, a height of 10 feet, and a length of slope of 40 feet, $E=\frac{L\times h}{l}=\frac{144\times 10}{40}=36$ pounds. It will be evident that though the required effort is halved when the length of slope is doubled, the total amount of work done ($E\times l$, or 36×40, or $18\times 80=$ 1,440 foot pounds) remains the same, and is thus independent of θ, the angle of inclination. Otherwise expressed, $E=L\sin\theta$, an equilibrium which an infinitesimal increase in E is sufficient to disturb.

ever, that in the process of multiplying an applied effort, no machine is able to give out more work than is put into it. The essence of the matter is that a weak force acting over a long distance can be converted into a strong force acting over a short distance, and *vice versa*. On the day that some unheard of device is prevailed upon to convert a weak force, acting over a short distance, into a strong force acting over a long distance, we shall be able to scrap all our motors and engines, and sit back and let this fabulous contrivance take over.

* The mathematics of this time-honoured procedure has since been investigated, and, neglecting friction, it transpires that $E\times l=L\times h$, where E is the effort, L is the load, h is the height, and l is the length of slope. This is merely a concise way of saying that the smaller the effort, the greater is the distance through which it must act for a given elevation to be attained.

In diverse forms and various combinations, the lever and the inclined plane make possible the bewildering array of complex mechanisms which function ceaselessly and uncomplainingly on our behalf to-day. But although the supreme importance of these devices can hardly be exaggerated, their widespread utilisation was dependent, among other things, upon advances in the realm of metallurgy—and the production of metal in quantity, and hence its extensive use in machine construction, is a comparatively recent phenomenon.

Throughout the so-called Ages of Stone, Bronze, and Iron, right down to the beginning of the 19th century, the material most useful to man was neither stone nor metal, but wood. Nearly all early tools and implements, including the wedge, the club, the bucket, the rake, and the hoe were made entirely of wood, as were the original versions of many of the more elaborate contrivances invented in later years, among them the wheel-and-axle, the ship, the crane, the windlass, the loom, and even the printing press. It is difficult to escape the conclusion, indeed, that if, by some whim of the high gods, the ancestral ape had been denied access to an arboreal abode, the descendants of primitive man would be primitive still.

Unlike timber, which was almost everywhere to be had for the taking, metals were much less readily come by. Though some of them are, in fact, plentiful enough, their main source is their ores, an unpromising assortment of oxides, sulphides, and the like. In consequence, most of them for long remained hidden, their very existence unsuspected. Of the 92 chemical elements from hydrogen to uranium, three-quarters are metals—yet fewer than a dozen of them were known to the ancients. Gold, which occurs native, together with silver and copper, have been prized since time immemorial. And iron beads, perhaps of meteoric origin, have been found in an Egyptian tomb dating back to about 5,000 B.C.,* while both lead and tin have also long been known. For the rest, only antimony, mercury, arsenic, bismuth, and zinc appear to have come to light before the dawn of the 18th

* More than 6,000 years later, the iron kitchen utensils of an English royal household were also classed among the jewels!

century.* Consider the following tabulation, which lists some now familiar examples in the order of their discovery, as presumed or known:

Prior to 18th century	During 18th century	During 19th century
Gold	Platinum (Wood, 1735)	Columbium (Hatchett, 1801)
Silver	Cobalt (Brandt, 1742)	Tantalum (Ekeberg, 1802)
Copper	Nickel (Cronstedt, 1751)	Osmium (Tennant, 1803)
Iron	Manganese (Scheele, 1774)	Iridium (Tennant, 1803)
Lead	Molybdenum (Scheele, 1778)	Palladium (Wollaston, 1804)
Tin	Tungsten (Scheele, 1781)	Sodium (Davy, 1808)
Antimony	Zirconium (Klaproth, 1789)	Potassium (Davy, 1808)
Mercury	Uranium (Klaproth, 1789)	Calcium (Davy, 1808)
Arsenic	Titanium (Gregor, 1791)	Magnesium (Davy, 1808)
Bismuth	Chromium (Vauquelin, 1797)	Cadmium (Stromeyer, 1817)
Zinc	Beryllium (Vauquelin, 1798)	Aluminium (Wöhler, 1827)

It will be observed that aluminium, an essential ingredient of clay, and the most abundant of all metals, was the last of those listed to be produced! Its preparation offered many difficulties, for its oxide (alumina) does not lend itself to ordinary reductive

* And of these, I. W. D. Hackh describes arsenic as a non-metal of metallic character, while antimony and bismuth are also not regarded as true metals by some authorities. All three substances, in fact, exhibit characteristics of both metals and non-metals, and they are sometimes relegated to an intermediate position, that of the *metalloids*. From this, it will be gathered that a concise and satisfactory definition of a metal is not easy to formulate, a matter in which it is certainly unwise to rely upon everyday experience. Thus, we are apt to think of metals in terms of solidity and hardness. But mercury is a liquid at all ordinary temperatures, while sodium is normally soft enough to be cut with a knife. Both, nevertheless, are unquestionably metals.

MASKEE *Sunday Dispatch*

That's Snagtooth—he invented fire!

methods, *e.g.*, heating with carbon. Success was finally achieved with the aid of metallic potassium, itself isolated but a few years before. Even so, aluminium remained little more than a laboratory curiosity until 1886, in which year Hall (and almost simultaneously Héroult) introduced the electrolytic process which remains the basis of modern production.

As for the time-honoured extraction of metals by smelting, the probability is that this was originally a chance discovery, and that a camp fire built on mineral-bearing stone resulted in a lump of crude copper or tin being found when the ashes were raked. Such an accident, however, would not as readily give rise to metallic iron, the production of which calls for a reducing atmosphere and a temperature of a relatively high order. But if the extraneous ore happened to be haematite or magnetite, *i.e.*, of particular richness and purity, and the fire itself hot and long-burning, continuously fanned the while by an obliging wind. . . .

However this may be, both wrought iron and steel have been known for 4,000 years or more, though it was not until the middle of the 14th century that a development of the famed Catalan forge gave rise to a blast furnace capable of actually melting iron ore, whereupon cast (or pig) iron became available in quantity for the first time. The reducing agent then in use was still the wood charcoal of old, and the ever-increasing demand for it led to a serious depletion of the forests—in England, during the reign of Elizabeth, there were no less than three Acts (dated 1543, 1558, and 1584) to restrict ironmasters from denuding the country of its natural raw material for the fleet. Then Dudley, in 1618, found a substitute for charred wood in coal, which Darby, almost a century later, in turn replaced with the much more suitable coke. It is this material which is still in use at the present time, and a modern blast furnace, towering 120 feet or so into the sky, can now produce 1,000 tons of cast iron a day, consuming in the process a like quantity of coke, 500 tons of limestone, 2,000 tons of ore, and 4,000 tons of air.

Curiously enough, pure (or at any rate comparatively pure) metals are but little used. It was long ago discovered that a mixture of two (or more) metals could give rise to an alloy possessing useful properties not exhibited individually by either (or any) of its constituents. Thus copper and tin, both relatively soft and weak metals, when mixed together in the proportion of 9:1, produce that tough and tenacious material called bronze. And to-day, the list of recognised alloys is a truly formidable one, many new names mingling with the old—alnico and brass, constantan and delta metal, duralumin and everdur, electron and german silver, gun metal and hiduminium, invar and manganin, monel metal and nichrome, osmiridium and permalloy, pewter and stellite. An outstanding example of how a metallic mixture can exhibit unusual and unexpected properties is provided by the Heusler alloys which, although composed of three nonmagnetic metals (aluminium, copper, and manganese) are themselves magnetic! No less remarkable are the fusible alloys, much used in safety plugs for steam boilers and automatic sprinkler devices. One such alloy contains bismuth, lead, tin, and cadmium in the proportions 4:2:1:1, and the melting points of these four components are respectively 271, 327, 232, and 321 degrees Centigrade. Yet, incredibly enough, the mixture itself melts at about 61 degrees—171 degrees C. below the melting point of its most fusible ingredient! And as if this were not enough, it can be used for the soldering of glass!

It is a moot point precisely when iron ceases to be iron, and becomes steel—essentially an alloy of iron and non-metallic carbon.* Even wrought iron, a comparatively pure form of the metal, is invariably associated with about 0·1 per cent of carbon, while in cast iron the proportion ranges from 2—5 per cent. The figure for steel, on the other hand, lies between 0·5 and 1·5 per cent., and the iron-to-steel transformation can thus be accomplished either by carbonising wrought iron, or by decarbonising cast iron. The ancients, who did not have access to supplies of cast iron, perforce adopted the first of these two methods. In this, known as the cementation process, wrought iron, packed with charcoal in clay containers, was strongly heated for a week or more, in the course of which treatment the requisite amount of carbon was gradually absorbed.

The mass production of steel followed the discovery by Kelly, in 1852, that molten pig-iron could be purified (and decarbonised) by forcing air through it, though it was not until 1914 that world output of the one surpassed that of the other (Fig. 4). In 1856, meanwhile, Bessemer devised a converter for carrying out the purification process, and subsequently acquired the Kelly

* There are, of course, many hundreds of special steels which, in addition to carbon, contain metals such as nickel, manganese, tungsten, and vanadium. These are the so-called alloy steels, thus named in order to distinguish them from the straight carbon varieties, which are designated non-alloy steels, *i.e.*, non-alloy alloys. Is there a nomenclator in the house?

II

Prime Movers and Power Sources

OF necessity, the early machines relied solely upon man-power for their operation, and vast hordes of slaves came to be dedicated to their motivation. How many man-hours, it may be asked, were devoted to quarrying and dressing the material needed for the Great Pyramid of Cheops? Or were expended in hauling the finished work up the earthen ramp (inclined plane) which led to the site? Or were required to complete the building of the tomb itself? We do not know the precise details of this monumental undertaking, but it has been estimated that more than 2,250,000 blocks of stone were used, some of them weighing as much as 15 tons apiece. No wonder report had it, in later years, that to complete the task, 100,000 men laboured at it for as long as 20 years!

Promise of ultimate relief from such drudgery (for all beasts of burden, including man) came with the revolutionary idea of driving machines by other machines, themselves harnessed to the inanimate powers of nature. History does not record the date of this momentous innovation, but the first practical engine of any consequence was in all probability the watermill, a device known to have been in use more than 2,000 years ago, and one which seems to have antedated the windmill by a millenium or so.

From pumping water and grinding corn, the new contrivances were put to sawing wood and spinning silk, to crushing ore and working metal, and towards the close of the 17th century, windmills of 10, and watermills of 20, horsepower* were in common

* A unit of power introduced by James Watt, and supposedly representing the average rate of working of a horse. It is the amount of power required to raise 550 pounds 1 foot in 1 second. A horse-power hour,

use. The most powerful prime mover in the world, according to Mumford, was then the Versailles waterworks: it developed 100 h.p., and could pump a million gallons of water a day to a height of more than 500 feet.

It was at this juncture that the steam engine, long foreshadowed by Hero of Alexandria (*circa* 150 B.C.), and more recently (if somewhat differently) conceived by Worcester, Papin, Savery, and Newcomen, finally emerged. It came, at long last, in answer to the insistent demand for a solution of the problem of removing flood water from the ever-increasing depths at which coal was being mined. The Newcomen version comprised a cylinder fitted with a piston, the downstroke of which was occasioned by spraying water into the steam-filled cylinder, whereupon condensation of the steam caused a vacuum. Essentially, the device was a pump (as, indeed, it was intended to be), and it remained for Watt to effect a series of improvements which led to the development of the steam engine proper.

The reciprocating steam engine, as thus evolved, remained without serious challenge (though not without considerable improvement at the hands of Trevithick, Hornblower, Woolf, and others) for a hundred years, during which period it was widely installed in an effort to meet industry's incessant calls for more power. And not only did it speed up production, but it revolutionised transport, heralding the doom of the sailing ship and the stage coach by giving rise to the steamboat and the railway locomotive. By the time Parson's rotary steam turbine appeared on the scene in 1884, with its promise of smoother running and higher efficiency, yet another prime mover was making its debut: the petrol engine of Gottfried Daimler, itself a descendant of the earlier gas engines of Otto, Langen, de Rochas, and Lenoir. An outstanding characteristic of the Daimler engine was its highly favourable power-weight ratio, a feature which was to prove of particular interest to advocates of the horseless carriage and to the votaries of heavier-than-air flight.

reasonably enough, is the amount of energy produced by an engine of 1 horse-power in the course of 1 hour—$60 \times 60 \times 550$, or 1,980,000 foot-pounds. This figure relates to the British horse-power; its Continental equivalent is somewhat less.

Wind and water power became of secondary importance as the heat engine was improved and developed, for this novel device provided a compact, self-contained, and mobile unit which, in one form or another, was eminently suitable for installation in mine and workshop, in ship and locomotive. But the heat engine was, and still remains, costly to operate, and its widespread use entails a prodigal expenditure of fuel, reserves of which are by no means unlimited. Moreover, of available fuels, only two are of any consequence. Between them, coal and oil provide most of the world's power, and it is significant that the production of these two vital commodities is characterised by ever-increasing mechanisation. Heat engines, that is to say, are fed by means of a host of secondary machines whose exertions they themselves promote.

In the U.S., the Carbide and Carbon Chemicals Company has devised and put into use an electrically operated coal getting machine specially designed to take advantage of exposed and horizontal seams. The mechanism, armed with a cutting head, burrows its way into the hillside trailing a control cable and a conveyor behind it. The coal is automatically brought out by the conveyor, the length of which is increased, as this becomes necessary, by a couple of operators stationed near the tunnel mouth. A third man directs the machine by remote control, steering it with the aid of signals which indicate whether it is cutting through hard rock (and hence veering off course) or soft coal.

In more conventional mining, particularly where the workings are ancient and tortuous, the introduction of modern labour-saving machinery presents many problems, and much ingenuity has been expended in devising safe and reliable cutting and loading mechanisms suitable for working where space and headroom are at a premium. Of these the Samson range of Mavor and Coulson may here be noted. It includes 19-inch, 15-inch, and 12-inch cutters (the dimensions are indicative of over-all height) —squat, sturdy, rectangular box-like machines, driven either by electric motors or compressed air turbines, whose endless chain-mounted picks ceaselessly claw their way into the coal face. No less effective is the MC 2 hydraulically operated loader, a

CLEW *600 Magazine*

low-lying, ramp-headed device fitted with two swivel arms which can scoop up loose coal at the rate of four tons a minute, and load it on to a belt, whence it is conveyed to a waiting skip for hauling to the surface. The hauling may itself be entirely automatic, the skip starting its journey as soon as it is loaded, increasing, maintaining, and retarding speed, before coming to a stop and discharging, entirely on its own accord. Thereafter, the coal is screened, washed, and dried in readiness for transportation and delivery—as likely as not to the hopper of some automatic stoker.

Although oil, like coal, is usually to be found underground, its extraction, once a field has been located, and a well drilled, is relatively easy, for gas pressure obligingly forces the fluid to the surface. And even when the pressure fails, as in time it must, the flow of oil can be assisted by forcing gas or air down the well, while as a last resort, pumps can be used. On the other hand, the crude substance thus obtained calls for a course of treatment far more extensive and elaborate than that required by freshly mined coal.

Crude oil is a mixture, the composition of which varies greatly, but its main constituents are always hydrocarbons of differing boiling point and varying degrees of complexity. Hence the first and decisive step in any refining process is distillation, in the course of which the crude material is heated to temperatures of up to 400 degrees C. in a pipe still, and then conveyed to a fractionating tower as a mixture of vapour and liquid. The liquid falls to the bottom of the tower, and the vapour condenses on trays at various levels, only the lightest constituent reaching the pinnacle. This topmost product is gaseous. Next comes motor spirit, and then kerosine, followed by gas oil, heavy gas oil, lubricating oil, and bottoms—fuel oil and bitumen.

Fractional distillation has been likened to running a quantity of gravel through a series of sieves, so as to separate it into groups of stones of different sizes. Nor need the sorting process stop at this. If, for instance, it were desired to increase the output of small stones, some of the larger ones could be broken up, and the debris so obtained re-sifted—and an equivalent procedure is, in fact, employed in the refining of oil.

The breaking up of large hydrocarbon molecules into smaller ones was first achieved with the aid of intense heat and pressure, and such thermal cracking is still conveniently applied to the low octane portion of the motor spirit which separates during the initial distillation. There results a high octane (anti-knock) spirit, a quantity of gas, and some fuel oil. A more recent process employs a catalyst, *i.e.*, a substance which causes other substances to undergo change without being changed itself. An

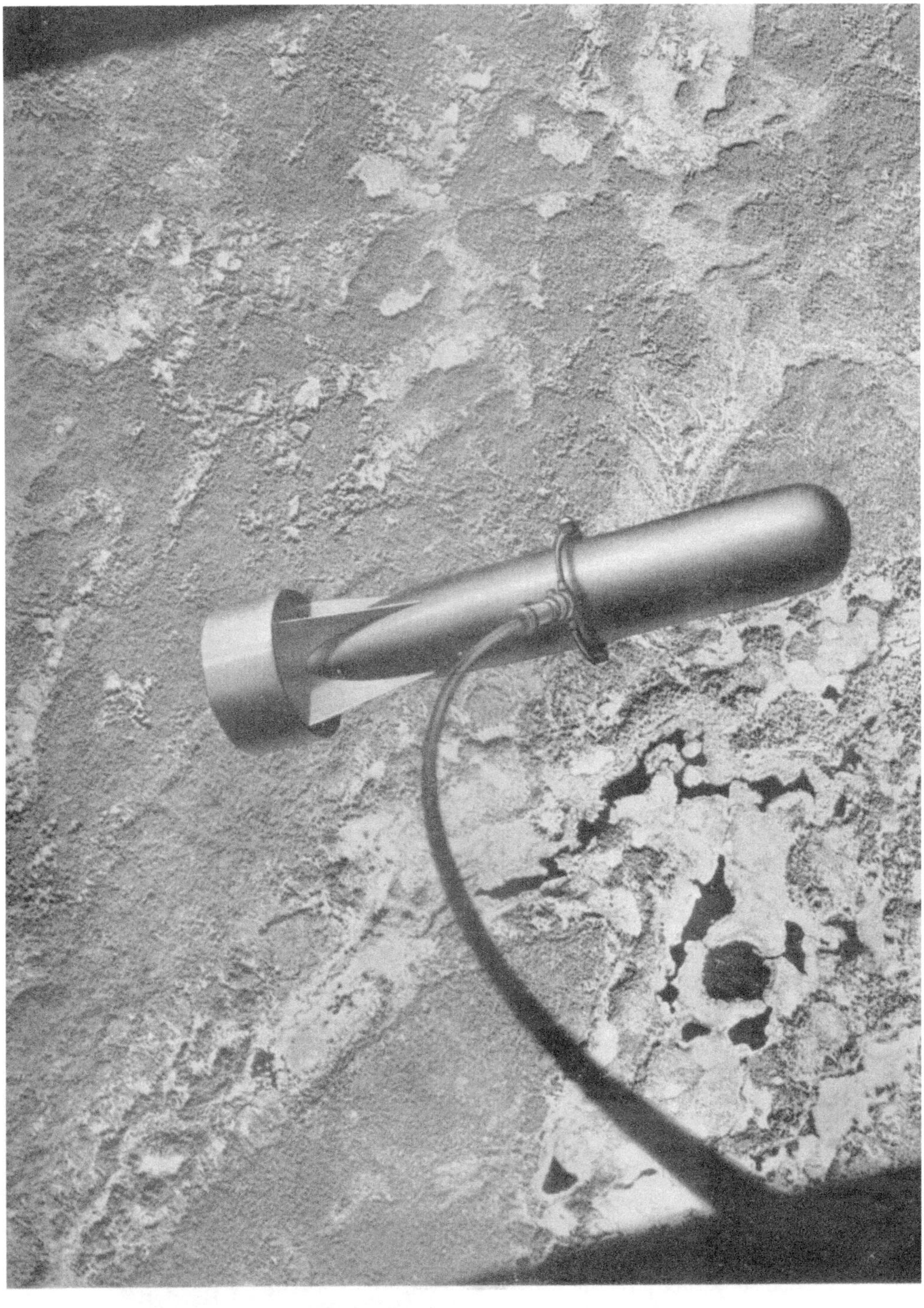

PLATE 2. An airborne magnetometer on a mineral survey over Canadian bush country. The instrument is sensitive to the earth's magnetic field, and reacts to any distortions in its regularity, such as would be occasioned by the presence of magnetic iron ore in the ground below. Photo: *Canadian Industries Ltd.*

wheel and the Francis turbine are well able to compete with the heat engine and its costly fuel requirements.

The only practical method of generating electricity on a worth while scale is by means of the dynamo, a machine which must itself be driven by an engine of one kind or another. Strictly, the dynamo does not generate electricity at all—it merely converts mechanical energy into electrical energy, which is then wire borne to where it is required, at which point it can be converted back again into mechanical energy by an electric motor. The great convenience of such a procedure needs no emphasis, and although the electric motor is not itself a prime mover, it is nevertheless a most important source of power. Indeed, the extent to which we have come to rely upon electrical energy, even in the home, at once becomes apparent when the supply of current fails, and we find ourselves suddenly plunged into darkness, or left half shaved, or faced with an underdone joint in the oven, or cut off from a radio concert, or glaring at a blank television screen instead of at a gratuitous intimation that normal service will be resumed as soon as possible. . . .

III

Elements of Control

IT will be evident that the watermill, so long as its source of power remains available, will keep on turning—subject, of course, to any necessary servicing and repairs. So, too, with the steam engine, in modern versions of which pressure is usually exerted alternately from above and below a movable piston, first forcing it up, and then forcing it down, control being exercised by a valve which automatically admits live steam to one side of the piston while exhausting spent steam from the other. Such an automatic cycle, however, was not a feature of the steam engine in its embryonic form. The original Newcomen pumping mechanism required the services of an attendant whose duty it was to operate two valves controlling the supply of steam. But a valve boy named Humphrey Potter (so the story goes), observing that his actions coincided with the never-ending rise and fall of an overhead rocking beam, tied cords to the beam, attached the other ends to the valves, and so induced the machine to look after itself.

Additional controls soon followed. Thus, cyclical variation of engine speed lends itself to continuous regulation by way of a flywheel, which acts as a reservoir, alternately absorbing and giving up energy during each stroke. Even so, should the engine be relieved of a considerable load, it will at once tend to race. But as the shaft increases its turning speed, it can be arranged for two rotating balls to be flung outwards, thereby bringing into play a valve which reduces the amount of steam entering the cylinder, so that engine speed returns to normal. In effect, the governor, without interfering with the basic operational cycle, enables the machine automatically to react to a sudden emergency, and to maintain control in the face of changing circumstances.

A somewhat more complicated, though hardly less familiar, sequence of operations is to be found in the automobile engine. Here, a four stroke cycle is continuously repeated in each of four or six or more cylinders, in the course of which a piston is pulled down from the top of its stroke by the rotation of a crankshaft. During this movement, an inlet valve opens, and an explosive mixture of petrol vapour and air is drawn into the cylinder; the inlet valve then closes. On its subsequent upward stroke, the piston compresses the explosive mixture, and a sparking plug is brought into action by means of a contact-breaker, itself controlled by the shaft. The mixture is thus ignited, and the resultant explosion forces down the piston in the one and only working stroke of the series. Thereafter, the piston starts to move upwards again, whereupon an exhaust valve opens, and the products of combustion make their escape. The valve then closes, and the cycle begins anew.

For many years, automatic controls remained a purely mechanical phenomenon, contrived with the aid of slide valves, cams, gears, levers, and the like. But in comparatively recent times, these have been supplemented by electro-mechanical and electronic contrivances, with the aid of which the operational capabilities of certain machines have been extended out of all recognition.

Devices which have contributed much to these revolutionary developments are the ordinary relay and the thermostat—the last-named an automatic temperature regulator often consisting of two narrow lengths of metal, having different coefficients of expansion, welded together. Such a bi-metal strip is remarkably sensitive to changes of temperature, and in spiral form its coiling and uncoiling can readily be made to operate a switch. As for the relay, this was originally used to promote long-distance telegraphic communication, and has since been incorporated in the automatic telephone, the calculating machine, and other such quasi intellects.

An outstanding feature of the relay is its supreme simplicity—essentially, it is nothing more than an electrically operated switch comprising a sensitive electromagnet, a spring-loaded armature, and a pair of contacts (Fig. 5). When energised by a flow of current, the electromagnet attracts the armature, moving it from the upper contact to the lower, and thus completing a local circuit. Immediately the flow of energising current ceases, the armature, no longer held by the electromagnet, is pulled back by the spring to its former position, and the local circuit is broken, so to remain until such time as the electromagnet is again

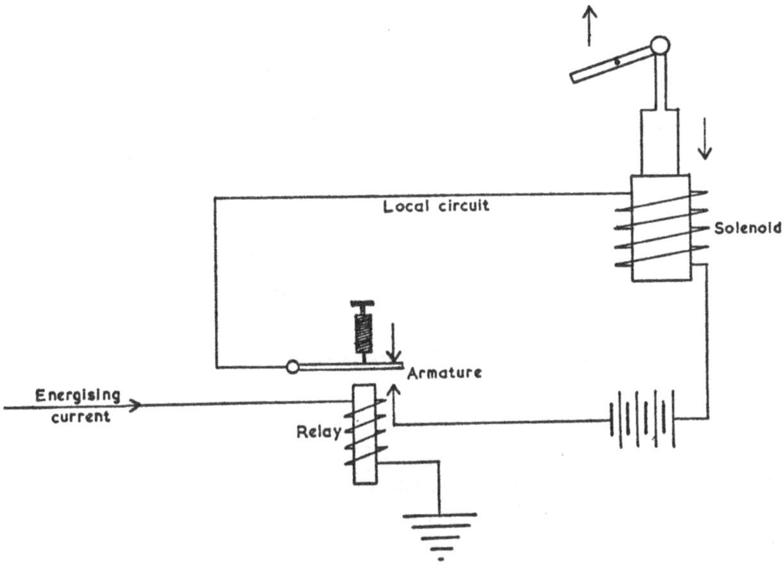

FIG. 5.—When current passes through the coils of the relay, a spring-loaded armature moves from the upper contact to the lower, thus completing a local circuit. A solenoid or other device can thus be energised.

energised. A similar contrivance, which has found widespread application, often in conjunction with the relay itself, is the solenoid (Fig. 5). This consists of a hollow, wire-wound cylinder, within which a magnetic field is produced when current flows through the surrounding coil. It is thus possible to arrange for an iron plunger to be drawn into the cylinder, and to interpret electrical impulses in terms of mechanical movement.

To-day, both the relay and the solenoid have been to some extent overshadowed, though by no means supplanted, by electronic devices such as the photo-electric cell and the thermionic valve. Of these, the photo-cell actually goes back to 1839, when Becquerel noticed that when one of the electrodes of a voltaic cell was illuminated, there was a change in the cell's electromotive force. The discovery, however, remained a scientific curiosity. Then some years later, in 1875, Willoughby Smith found that the electrical resistance of metallic selenium was altered by a beam of light. But once again, no explanation was forthcoming. And in 1877, Hallwachs, following Hertz, observed that a zinc plate became charged under the action of ultra violet light. The reason was unknown—until, at the beginning of the present century, Lenard proved that it was due to the expulsion of electrons. To-day, all three discoveries are recognised as different manifestations of an identical phenomenon: the photo-electric effect, *i.e.*, the discharge of electrons from a metal surface under the influence of light. Subsequently, it transpired that although all metals are so affected, only those of the alkali groups (such as sodium and potassium) are sensitive to visible light.

Photo-voltaic (copper-cuprous oxide) and photo-conductive (selenium) cells have found a variety of uses, the one in light measuring instruments (*e.g.*, the photometer) and as a rectifier, and the other in applications such as the control of street lamps, which it automatically turns on at dusk, and faithfully extinguishes at dawn. The selenium cell, however, has to a large extent been superseded by the photo-emission cell, which finds ever-increasing application. The device consists of two electrodes, an alkali metal-coated cathode and an adjacent anode, sealed in an evacuated bulb. When light falls on the cathode, streams of electrons are emitted, and these (by arrangement) are attracted to the anode. In effect, there is a light-stimulated flow of current which, after amplification, can be made to operate all manner of controls. These controls, moreover, can be so contrived that they function either when light falls on the cell, or when an illuminating beam, not necessarily of visible light, is cut off from it. Thus an approaching vehicle can be induced to open its own garage doors by means of its headlights, or merely by interrupting a beam of invisible light directed across its path. Similarly, the device has been used to relieve a driver of the necessity of constantly dipping his headlights on encountering oncoming traffic at night. Again, it has been used to provide an unseen fence round dangerous machinery, so as to prevent the descent of a cutting blade if the hands of an operator are not safely out of the way. Or it can pick out unlabelled articles from labelled ones. Or give protection to premises by warning of some unauthorised intrusion. Or signal train arrivals and departures. Or control the opening and closing of lift and other gates. In short, its applications are as numerous as they are varied.

Like the photo-cell, the thermionic valve had its beginnings in an effect, noticed by and named after Edison, for which no explanation was immediately forthcoming—that an electrode in the vicinity of a glowing filament acquired a negative charge. It has since been shown that any electrical conductor, if it be heated sufficiently, will emit electrons, much as boiling water gives off steam. And as a result, the surrounding space becomes negatively charged, a fact which may readily be demonstrated by means of a red hot poker and a gold leaf electroscope.

The thermionic valve was born when Fleming, at the beginning of the present century, devised an electric light (vacuum) bulb with a difference: in addition to the conventional filament, which emitted light and heat (and electrons) when the current was switched on, he inserted an anode in the shape of another electrode. Now in the ordinary course of events, electrons leaving a heated filamentary cathode remain in the vicinity of it as a negative 'space cloud.' For by their loss, the filament acquires a positive charge, and thus has an attraction for the hovering electrons, most of which will normally return whence they came. But if the anode is maintained at a sufficiently high positive potential, the ambient electrons will be drawn there instead, and a one way flow of current will thus be established.

From the diode (or two electrode) valve, capable only of rectification, Lee de Forest evolved the triode, or amplifying, valve. He interposed a third electrode (the grid) between cathode and anode, and thereby introduced a means of voltage control. This was in 1907, and the process of electrode addition has continued ever since. The tetrode, with a second grid, made its

BROCKBANK *The Motor*

Madam's bug-deflector is perfectly sound, but madam has fitted it on backwards.

appearance in 1916, and ten years later the pentode followed in its wake. Hexodes, heptodes, and octodes, boasting four, five, and six grids respectively, have also been produced.

The diode valve has found use pre-eminently as a device for the rectification of small radio frequency voltages, but larger versions of it may be employed to obtain direct current for industrial purposes from A.C. mains supplies. In a vacuum tube, however, the electrons, because of mutual repulsion, impede one another in their journey from cathode to anode, an adverse effect which in power rectification may require the maintenance of a voltage drop amounting to several hundreds of volts across the valve. This can be avoided by the use of valves containing vapour (mercury) or gas (such as hydrogen, helium, or argon), which becomes ionised by the electron stream, and so gives rise to positively charged ions. The negative space charge is thus neutralised, and a voltage drop of some 20 volts only is sustained. As an added refinement, two (or more) anodes may be used, and so arranged that they carry current in turn, whereby full-wave rectification can be achieved. And as with the radio version, by the addition of one or more grids, the valve becomes a controlled rectifier. It is then known as a thyratron.

At the other end of the scale, remarkable and far-reaching

developments have taken place in recent years in the sphere of semiconductor elements. These have long been known to possess rectifying properties, and indeed certain of them played an important part in the early days of radio, when extensive use was made of galena (lead sulphide) and other crystal detectors. In the 1920's, however, crystal detectors fell into general disuse and were replaced by the thermionic valve, which remained unchallenged and supreme for the next twenty years. But during the last world war, the conventional vacuum tubes were found wanting in connection with the rectification of very high frequency radar signals, and a solution of the problem was sought among semiconducting materials, particular attention being paid to silicon and germanium. The immediate outcome was the germanium diode, a tiny crystal sprouting two wires, able to perform the function of a vacuum diode, and yet possessing the very considerable advantages of comparative smallness, shockproof construction, and no heating filament (with its inevitable demand for ancillary power supplies)—in short, a revolutionary device of immense significance, particularly of military significance, involving not only radar, but the entire field of communications and control, from rocket guidance and aircraft equipment to walkie-talkies and gun-directing electronic computers.

The germanium diode, moreover, was but a beginning. In June, 1948, Bell Telephone Laboratories announced the transistor, or germanium triode—an equally small and effective amplifying device of the point-contact type. Nor was this all. Less than two years later it was followed by the photo-transistor, a miniature germanium equivalent of the photo-electric cell, and in 1952 the tetrode transistor made its appearance, endowed with a fourth electrode. To date, some 40 different types have been developed by Bell, perhaps the most important of which is the junction (type) transistor of Dr. William Shockley, the first news of which was given in July, 1951. It consists of a tiny sandwich of germanium, so treated that its three layers have different electrical properties. Two types of germanium go to make up the sandwich—*n*-type (n = negative) and *p*-type (p = positive). The *n*-type is produced by adding to the germanium a trace of an impurity such as arsenic, able to provide an abundance of electrons.

The *p*-type, on the other hand, contains a trace of a substance having a deficiency of electrons. The normal sandwich arrangement is either *n-p-n* or *p-n-p*, with connecting wires touching each layer, and the complimentary properties of the two arrangements have led to their combined use in many novel circuits, such as the unique RCA audio amplifier consisting of nothing but four transistors, two of each kind.

The new semiconductors made their appearance, appropriately enough, at a time when the American Armed Services were calling for more and more to be packed into less and less—the ideal being, as RCA engineer D. F. Schmitt has wryly remarked, that electronic equipment should occupy no space, and have zero weight. To this end, the miniaturisation of radio and allied components was followed by their sub-miniaturisation, and wires and coils of three dimensions made way for two dimensional printed circuits, this in itself a far-reaching innovation of great promise. The overall dimension of the conventional valve, meanwhile, which had already decreased from four inches (1920 version) to three, was further reduced to two, and ultimately reached one and a half, all the while undergoing a proportionate reduction in diameter. A complete transistorisation of valve circuits may well carry the process to its logical conclusion: the germanium triode occupies about 1/400th of a cubic inch!

PART TWO

Automata in Action

★

I

Mechanisms Emergent

AFTER a gestative period of some 6,000 years, an automatic ribbon loom is said to have been produced in Danzig in 1579, or at any rate not later than 1661. The event, however, would not appear to have been regarded as a particularly happy one at the time, for the story is that the City Authorities, fearful of the unemployment which would be caused among the weaving fraternity, had the luckless inventor quietly done to death, and his unwelcome offspring suppressed. Be this as it may, in the years which followed, the forbidden device was to be found busily at work in many parts of Europe. And in a perfected version of the draw loom, subsequently devised by Jacquard, the lifting of the warp threads was automatically contrived by means of rods and punched cards. Those rods whose ends did not pass through the holes were lifted, and a repetitive pattern could be produced merely by the use and re-use of a particular set of cards.

A woven fabric has been succinctly defined* as the result of interlacing a single weft end of yarn over and under a number of warp ends running in a direction at right angles to the weft. To this end, shedding (whereby alternate warp threads are raised so as to form an alley) is followed by picking (in which the weft is carried across the warp) and battening (so that the weft is compacted and the fabric made firm). In early days, the picking operation was performed by hand with the aid of a stick with a hooked end, a laborious undertaking which was eventually speeded up by the throwing of a shuttle between two weavers, from one side of the loom to the other, through the alley aforementioned. The need for two operatives vanished with

* By H. de G. Gaudin, in a Paper, *The Case For Conversion To Automatic Weaving*, given at the Cotton Board Conference, Harrogate, 1948.

the introduction of Kay's flying shuttle in 1733, a mechanical device which enabled one man to effect the to and fro movement of the shuttle by himself.

Output was so increased (by something like a hundredfold) that the demand for yarn threatened to exceed the supply, an emergency which was answered, first in 1767 by Hargreave's spinning jenny (a hand-operated contrivance which featured a multiplicity of spindles) and shortly afterwards by Arkwright's water frame (a power-operated machine which made use of the roller principle of Wyatt and Paul)—and some ten years later by Crompton's spinning mule (so-called because it combined the salient features of its two predecessors). The immediate outcome of this mechanisation was not only thread in much greater quantity, but thread so much improved in quality that it was sufficiently strong to make the warp—hitherto, cotton had been used only for the weft. This, in turn, led to an increased demand for the raw material, and to the invention of a machine for separating cotton from its seeds (Whitney, 1793).

The question of a power-operated loom, meanwhile, had been receiving attention, and the unsatisfactory Cartwright version which made its appearance in 1785 led eventually to Horrock's much improved design a quarter of a century later. This necessary formality having been observed, the emergence of a fully automatic loom was merely a question of time. Among the subsidiary problems which remained to be overcome were the automatic replenishment of the weft supply (which called for a self-threading shuttle, into which a full bobbin could be inserted without stopping the loom); some means of taking care of the two ends of the weft (a special cutter was evolved); and a method of supplying full bobbins to the shuttle (a magazine feed provided the answer). Success—in the shape of the Northrop machine—was finally achieved in 1894, in which year an installation was made at the Queen City Cotton Mills in Burlington, Vermont, in the U.S. Here, for the first time in the long history of weaving, one operative attended to no less than sixteen looms—a figure which has since risen to eighty or more.

The discovery, in the eighteen eighties, of how to make a silk substitute by artificial means opened up possibilities which have

WIMSEY *600 Magazine*

since been exploited to the full. Of the four methods of manufacture which were finally evolved, the viscose process proved to be the most economical. The raw material is wood pulp (cellulose), and after bleaching it is soaked in caustic soda solution and then treated with carbon bi-sulphide. A gelatinous mass results, and this, dissolved in caustic soda solution, is forced through fine jets into an acid medium, which neutralises the alkali and reprecipitates the cellulose as a fine, silk-like thread.

In 1938, after long research, the Industrial Rayon Corporation of America put into operation a continuous viscose spinning machine in which, untouched by hand, an unending length of yarn is spun and processed, so that it emerges desulphurised, bleached, washed, dried, twisted, and ready for use. Travelling through the machine from each jet to the bobbin of finished material is a length of thread which stretches for more than a quarter of a mile—yet the entire process requires less than six minutes to complete!

Like the beginnings of spinning and weaving, the origin of glass making is also lost in antiquity—Flinders Petrie has described a moulded amulet which dates back for some nine thousand years. And the technique of glass blowing, though less venerable, is likewise an ancient one. Supposedly developed by the Phoenicians about 300 B.C. (or by the Egyptians even earlier), it has for centuries been carried out with the aid of a hollow iron tube and a strong pair of lungs, a blob of molten glass being shaped by a combination of measured blowing and skilful pipe twirling. In due course, repetition work was facilitated by the use of iron moulds, wherein the expanding glass was impressed, and attempts to mechanise the process produced a glass blowing machine in 1886. Twenty years later, the famous automatic bottle-making machine of Michael Joseph Owen made its debut.

In modern versions of this intricate assemblage, a cycle of forming operations is carried out by ten or fifteen heads, each head having its own mould and performing the necessary sequence of operations. In the course of these, a 'vacuum and blow' process is employed—molten glass from a revolving pot is sucked into a mould, and blown into shape by compressed air after its connection with the glass supply has been severed. As many as 100 bottles a minute can be produced by the fifteen head model.

Other types of machine, such as the Hartford-Empire I.S., work on what is termed the 'blow and blow' principle, and operate in conjunction with a feeding mechanism which automatically shapes, cuts, and delivers gobs of glass to it, *i.e.*, glass in measured quantities having the right size, weight, and temperature. The I.S. machine is made up of a number of Individual Sections (five is standard), each complete in itself, comprising one blank mould and one blow mould. When a gob of glass enters a blank mould, it drops to the bottom and settles uniformly into a neck ring. Air pressure is then applied to complete the process. This, the settle blow, is immediately followed by a

PLATE 4(*a*). A multi-bulb mercury-arc rectifier installation. It provides a D.C. power load of 1,000 kW. Photo: *The Electric Construction Co. Ltd.* (*b*). Electronic control assembly of the TOCCOtron 20 induction heating machine for the automatic hardening, brazing, or annealing of small parts. Photo: *The Ohio Crankshaft Co.*

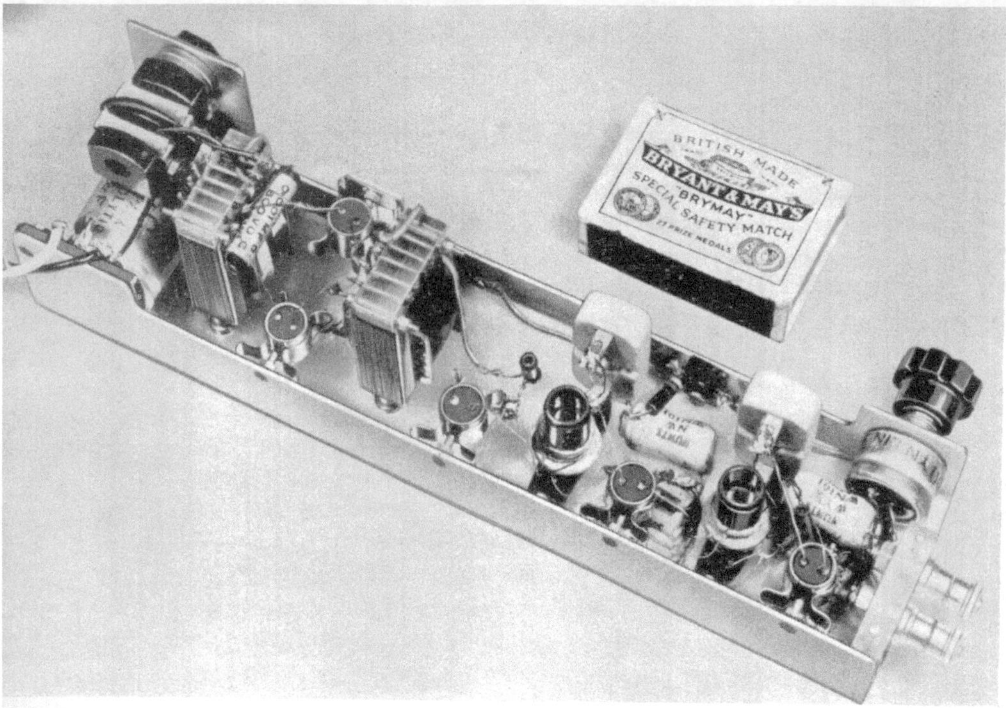

PLATE 5(*a*). Miniaturisation of a radio receiver, achieved with the aid of five germanium triodes. Photo: *The General Electric Co. Ltd.* (*b*). The section of an electronic adding unit shown in the foreground contains five transistors in place of the four valves and four transformers which go to make up the equivalent piece of apparatus in the background. The use of transistors saves considerable space, and power consumption is reduced by three-quarters. Photo: *Radio Corporation of America.*

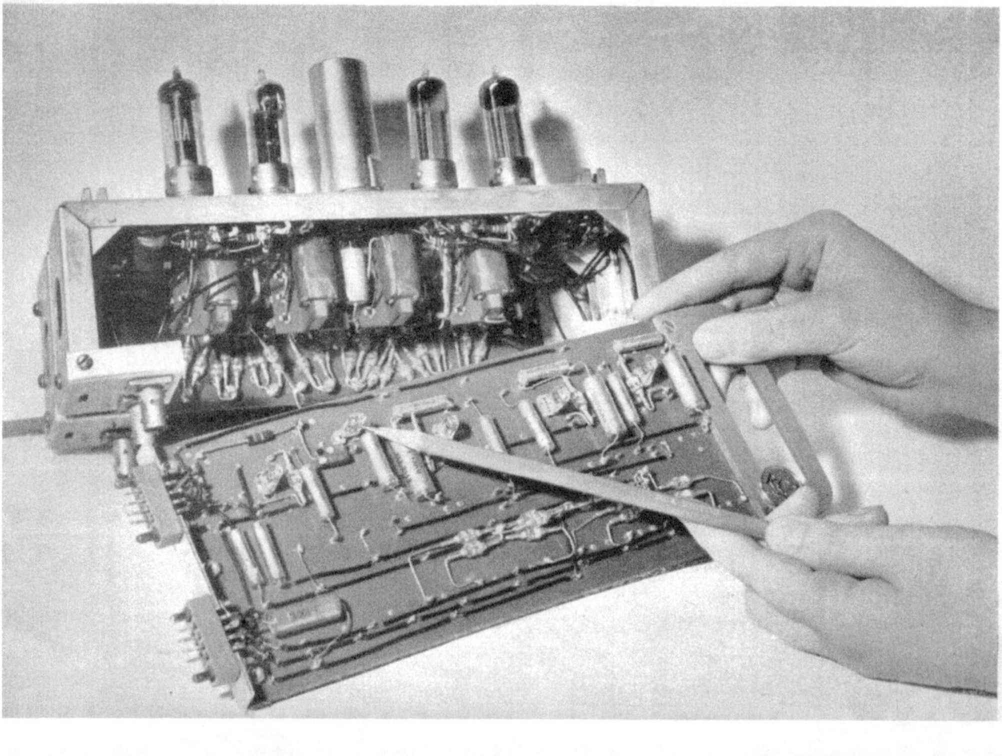

PLATE 6. An installation of 150 automatic lathes (photographed while working). Photo: *Usines Tornos S.A.*

PLATE 7(*a*). The new Lynch '10' automatic glass forming machine. There are two revolving tables, with six blank moulds on one, and six blow moulds on the other. The machine can produce some 70,000 half ounce containers in the course of a twenty-four hour working day. Photo: *British Hartford-Fairmont Ltd.* (*b*). Close-up of a modern Owens bottle machine. The glare of molten glass in the revolving pot from which the machine is fed can be seen in the background. Photo: *Owens-Illinois Glass Co.*

counter blow from the mouth end, the effect being to hollow out the blob of glass. The parison (partly shaped blank of glass), held by the mouth, is then inverted by a mechanical arm and transferred to a blow mould. Here, after a reheat, it is subjected to a final blow of air, and so made to assume the shape of the mould. The complete cycle requires from 5–15 seconds, the exact length of time depending upon the size of the product.

Until about twenty-five years ago, glass bulbs for the electrical industry were all made by hand. Then the process was completely revolutionised by the appearance of the so-called ribbon machine, a development sponsored by the Corning Glass Works in the U.S.* The process is as follows: from a furnace, in operation day and night, molten glass pours into the machine in a continuous, white-hot stream, falling vertically between two water-cooled rollers. It emerges, horizontally and at right angles, as a fast moving ribbon, plain on the underside, but bearing a series of evenly-spaced blobs on the other. These blobs—bulbs in embryo—are formed continuously by indentations on the surface of one of the rollers, and their distance apart is governed by the size of the bulbs to be made. Normal centre-to-centre spacing is between three and four inches.

Bulb formation is the result of a rapid succession of operations, all perfectly timed and automatically executed by the machine. The ribbon of hot glass travels on a moving platform made up of an endless chain of orifice plates—flat, rectangular pieces of metal, each with a circular hole in its middle, evenly spaced at three inches or more, as the case may be. Running directly above the ribbon of glass, on an endless belt resembling that of a tractor track, is a line of blowheads, each one of which positions itself in turn over a glass blob. And no sooner is one of the blowheads thus aligned than a puff of air issues from it, forcing the glass blob down and through the hole in the orifice plate, and thus starting the process of bulb formation—blowhead, glass ribbon, and orifice plate all the while moving along in unison.

* Two specimens of this amazing machine have recently, if somewhat belatedly, reached the British Isles. They are installed—as a joint G.E.C.-B.T.H. enterprise—at the factory of Glass Bulbs, Limited, Harworth, where the author was privileged to watch one of them in operation.

Underneath this procession is another endless track, similar to that carrying the blowheads above, on which are mounted cork-lined metal moulds, split from top to bottom, and hinged at the base. Their movement, too, is in perfect step with the ribbon of glass overhead, and before the steadily growing bulb can become too large and distended, it is enveloped by an approaching mould, the two open jaws of which automatically close over it, and then begin rapidly to spin. The cork lining, carbonised by prior usage, and earlier in the circuit sprayed with water, at once checks, shapes, and polishes the expanding glass. Sticking in the mould is prevented by the cushion effect of clouds of steam which arise when the hot glass encounters the damp lining, and which make their escape through a multitude of tiny holes which pierce both lining and mould.

As soon as the formation of a bulb is complete, its attendant blowhead and mould simultaneously go their separate ways—the one turning upwards and away, and the other downwards—while the bulb, one of a line of globes left dangling from the parent strip, carries straight on, first to encounter cooling jets of air, and then to receive a smart tap from a hammer, which unerringly detaches bulbs at a rate of up to 700 a minute as they emerge from the moulds in a never-ending stream. Each bulb falls neatly into a scoop attached to a revolving turntable, and is carried round until it is over a moving conveyor belt. Here, eight at a time, the scoops deposit their load, and then go round for more, while a continuous flow of bulbs is transported by the conveyor, *via* an annealing oven, to the packing station.

Just as glass blowing was of necessity preceded by the discovery of glass itself, so the development of printing (as opposed to writing on papyrus, or on clay or wax tablets) had to await the invention of paper. This momentous occurrence appears to have taken place in China in pre-Confucian times, and it was in the East, too, that the art of printing by means of wooden blocks subsequently arose. The first movable type (also made of wood) has been attributed to Pi Sheng, *circa* 1041, and some four centuries later, the innovation appeared in Europe—independently evolved by a Dutchman, one Laurens Janszoon, *alias* the Coster, according to some authorities, and by Gutenberg and Shoeffer, of Mainz, according to others. Either way, movable wooden type was soon replaced by metal type, cast in a mould—an improvement which had already been anticipated in Korea by a hundred years or more.

Until the middle of the 19th century, all printing was done on a flat-bed machine, essentially a level surface on which a frame containing type was placed face upwards. The type having been inked, paper was then lowered onto it by a folding attachment, and the requisite impression obtained by the application of pressure.

Stereotypy (the fabricating of solid casts in type-metal from pages of movable type) and lithography (the art of writing or engraving on stone, and printing therefrom), devised respectively by Ged in 1725 and Senefelder in 1796, were followed by Koenig's steam-operated press in 1811. In this, although the type bed remained flat as in the conventional hand machine, the paper was pressed onto the type face by a revolving cylinder. Output was thereby considerably increased, and by 1814 the Koenig press was producing copies of *The Times* at the unheard of rate of 1,000 sheets an hour.

In 1866, the American R. M. Hoe devised a rotary press in which the type itself was placed on the cylinder, and twenty years later MacDonald and Calverley in England patented a press (subsequently named after J. Walter, of *The Times*) in which the cylinder type was replaced by curved stereotype plates, an arrangement well suited to newspaper production, as an unlimited number of copies could be printed in a given time by means of additional machines and duplicate plates. But although the stereotype process enabled type-setting to be reduced to a minimum, it did not abolish it entirely, and the need first to set the type by hand was a constant brake on production. Despite many attempts to solve the problem, it remained for Ottmar Mergenthaler, in 1886, to provide the answer with his Linotype machine, a contrivance of the utmost ingenuity, and one which, true to its name, actually produced lines of type.

The traditional method of type-setting entails a laborious selecting of letters by hand, one at a time from a type case, and a painstaking building up of words and sentences into lines and columns. Mergenthaler's machine not only does all this—it also makes its own type as it goes along. Banks of lettered keys, not unlike (though more numerous than) those of a typewriter, release a small brass mould as each key is depressed. These moulds (matrices) travel down a chute in ordered procession, and form a line not exceeding the prescribed width, the final spacing being achieved by means of steel wedges which automatically drop into place between the words, so forcing them to occupy the full width of the column. From a melting-pot, gas fired, and fitted with a pump, molten type-metal runs into the assembled moulds, forming on solidification a bar (known as a slug) of type. Once this has been cast, an elevating device conveys the moulds to a distributing bar, which sorts them out and returns each to its own particular magazine, there to await re-use. The line of type, meanwhile, passes on to a metal tray (galley), and so takes its place in the column which is quickly being formed.

The Linotype machine was followed almost immediately (in 1887) by Tolbert Lanston's Monotype, a mechanism which, as it name suggests, casts and sets letters individually, an arrangement which has the advantage that the correcting of a one-letter error can readily be done by hand from a 'sorts' case, whereas with the Linotype even a minor alteration entails the re-casting of an entire line. Each machine, however, has its own particular merits, and while the Monotype is favoured in book printing, the Linotype is still preferred for newspaper work. With its aid, compositors can now keep pace with the demands of the pressroom, where a modern Hoe machine, fitted with an automatic folding attachment such as that long ago devised by Duncan and Wilson, can print and deliver folded papers in ordered bundles at the rate of some 200,000 copies an hour—a necessary achievement when it is remembered that some dailies boast circulation figures which run into several millions.

The whole of modern industrial production rests, at base, on a relatively small group of machines which are used in the making of all other machines, including replicas of themselves—the so-called machine-tools. In as much as nearly all tools *are* machines, these indispensable mechanisms are somewhat inaptly named.

There is, moreover, a noticeable lack of agreement as to precisely what a machine-tool is, and what it is not. The definition of one authority includes the power-driven tools of the woodworker, whereas that of another specifically excludes them. A third restricts the term to certain metal-cutting machines, and adds the proviso that they must be non-portable, a requirement which is conditionally waived by a fourth, albeit both are in agreement that a motor-driven hand drill, though indisputably a machine, and unquestionably a tool, is nevertheless not to be regarded as a combination of the two. In short, confusion reigns, and anomalies abound.*

By general consent, however, the father of all machine-tools is acknowledged to be the lathe, which in its original form was nothing more than a simple turning device, made entirely of wood, and motivated by the pulling of a cord wound round a spindle. Eventually, bent sapling and cord gave way to the foot treadle, and the foot treadle was ousted by steam power and belting, later destined to be supplanted in turn by the built-in electric motor. In the meantime, a slide rest was added, and metal construction replaced that of wood. Thus, in modern guise, emerged that erstwhile mechanical maid of all work, the centre lathe—essentially a pair of pointed supports on which the workpiece is rotated, and a tool rest by means of which a cutting tool can be firmly held or fixed.

Various modifications and refinements followed. The introduction (in the U.S.) of the chuck, a gripping device for holding and rotating work unsuited for mounting between centres, was one such elaboration. An allied innovation was the turret or multi-tool contrivance, in effect a revolving head enabling one tool after another to be brought into use in quick succession.

* According to an article in the 1938 edition of *Van Nostrand's Scientific Encyclopedia*, the machine-tools recognised as such by the Machine Tool Builders' Association in the U.S. are limited to the following: Boring machines; Boring and turning mills; Broaching machines; Cut-off machines; Drills; Gear cutters; Grinders; Milling machines; Hobbing machines; Planers; Key seaters; Rack-cutting machines; Screw machines; Shapers and slotters; Tapping machines; and Lathes.

The reference in the 1947 edition of this work, however, is less specific.

Other advances, meanwhile, led directly to the development of entirely new mechanisms—offshoots such as the boring machine, the drilling machine, and the grinding machine.

This mutative process was initiated by the English toolmakers of the Boulton and Watt era, in an attempt to meet the more exacting demands of would-be precision engineering. The call for steam boilers exhibiting a reasonable uniformity of diameter throughout their length was answered by ironmaster John Wilkinson, with his improved version of the Smeaton boring machine. Henry Maudslay, meticulous as always, perfected both the slide rest and the screw-cutting lathe. And the hydraulic press, not to mention a planer (itself the ancestor of the shaper and the slotting machine) was introduced by Joseph Bramah, the versatile Yorkshireman whose creative genius had already produced that porcelain monument to Anglo-Saxon fastidiousness, the water closet.

The one outstanding feature of modern machine-tool design is the trend towards self-operation. Thus, among the Heald Machine Company's range of standard grinding machines, there are models which require nothing of the operator except that he maintain the level of the loading chutes, and occasionally change a worn wheel: loading, grinding, sizing, and unloading are all automatically performed. And at the other end of the scale, the Wean Equipment Corporation recently designed what is claimed to be the largest multiple head automatic grinder ever built, a 350-ton monster endowed with a control system containing 15 miles of wire and 60,000 electrical connections, capable of handling the hulls of military tanks. Each of its ten grinding heads is independently controlled, and entirely self-operative. As soon as the required depth of cut has been reached, the circular motion of the abrasive wheel stops; the grinding head withdraws on its own accord, and cleans itself in preparation for the next operation.

Again, machines which cut steel bars to length, such as the Racine hydraulically controlled models, require no further attendance once the stock has been loaded and the gauge set to the desired length. Thereafter, all operations are taken over by the machine, from the pushing up of stock and the lowering and raising of the saw blades to the opening and closing of the holding vices. Similarly, the complexities of spiral bevel gear cutting can also be performed without human intervention. In the Oerlikon Spiromatic No. 2 machine, three continuous rotary motions are simultaneously employed in the process: rotation of the cutter; rotation of the workpiece; and the generating roll of the cutter relative to the work. Control is electro-hydraulic, and once the wheel blank is in place, the working cycle starts at the touch of a

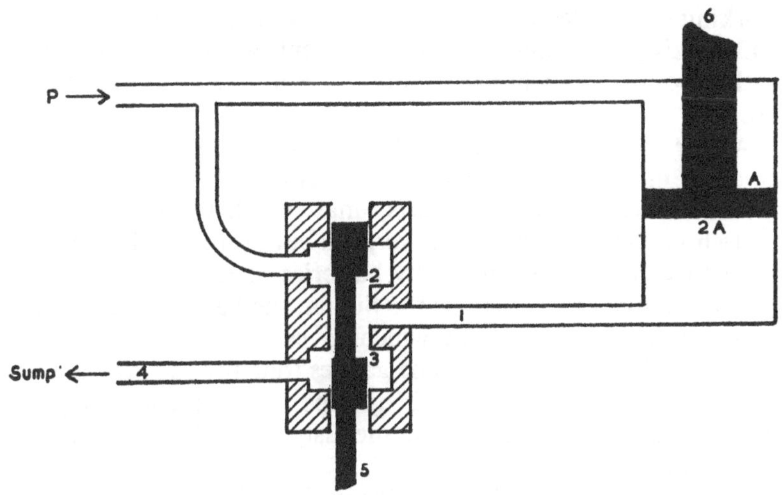

FIG. 6.—The Metrovick patented hydraulic control system. A pilot valve (5) acts as a hydraulic potentiometer to vary the pressure in pipe 1. In the position shown, openings 2 and 3 are equal, and the continuous flow of oil through the valve and back to the sump *via* pipe 4 causes half the supply pressure P to drop across opening 2, and half across opening 3. The piston 6 is so designed that if the effective area on the top side is A, the area on the bottom side is $2A$. Thus the pressure on the top side is $P \times A = PA$, and that on the bottom side $P/2 \times 2A = PA$, *i.e.*, the forces are equal, and the piston is at rest.

If the pilot valve member 5 is displaced upwards, opening 2 is increased, and opening 3 is reduced, by equal amounts. As a result, the pressure in pipe 1 increases, and piston 6 moves upwards. Conversely, a downward displacement of member 5 decreases the pressure in pipe 1, and piston 6 moves downwards.

lever, whereafter the machine carries on to completion, finally indicating that its task is done by coming to a stop.

Certain standard machines, for example boring mills and planers, can be fitted with profiling equipment, hydraulically controlled (Fig. 6), which enables them faithfully to reproduce the contours of a master template or sample workpiece. Other mechanisms, such as oxygen cutters, are amenable to control by a tracing head which automatically follows the outline of a drawing,* while Bell Telephone Laboratories are reportedly working on an electronic device, which will set about designing and making drawings of new equipment, when it is presented with an outline of what the proposed equipment is required to do.

As for the lathe, fully automatic versions of it are now available in almost endless variety, and many models are designed to turn out small component parts, unattended, by the thousand every hour. These parts, moreover, can be automatically checked as fast as they are produced by batteries of multi-dimensional inspection machines, such as the Sigma, the basic unit of which is an electric signal gauge which gives results to within an accuracy of some 0·00005″. The gauges (one for each feature to be checked) have been used in groups to inspect as many as twenty-eight dimensions simultaneously, though half a dozen or so is more usual.

The significance of this ever-increasing trend towards automatism among the machines which make the machines will be evident, and it will be pursued to its logical, if somewhat startling, conclusion in due course.

* The arrangement is such that a spot of light, exactly bisected by the edge of the line that is to be followed, is held in this position by a steering motor; any deviation immediately causes a control to turn a travel motor so as to restore the spot to its position.

II

Robots In Our Midst

FOR better or for worse, we are now surrounded by a multitude of self-operating mechanisms, ranging from simple domestic appliances to industrial equipment of bewildering complexity, the existence of most of which we have come to take largely for granted. In many homes, one or more unobtrusive calculating machines diligently record the amount of gas or electricity which is being consumed by a small army of other resident mechanisms, some of them no less automatic in operation than the meters themselves. Thus, water and space heaters silently maintain such temperatures as are desired; electric toasters, similarly thermostatically controlled, not only switch themselves off, but eject their contents as well; radiograms automatically change their quota of records; and dishwashing and other machines unerringly undertake a succession of spraying, washing, rinsing, and drying operations.* There is even a gadget which will prepare an early morning cup of tea at any designated hour, first boiling a quantity of water, then siphoning it into a pot, and finally announcing that the infusion is ready for consumption by ringing an alarm. But perhaps the most ingenious of all household contrivances is the noiseless refrigerator of the Electrolux type, for not only does it contain no moving parts, but it refrigerates, improbably enough, by the application of heat!

This astonishing contrivance is made up of four main components—a boiler, a condenser, an evaporator, and an absorber,

* An American friend reports a clothes washing machine which first weighs the load, then measures out the correct quantity of detergent, and thereafter continues with its ablutionary measures until the entire wash has been properly cleaned. An automatic clothes drier then takes over, carrying out its functions until everything is thoroughly aired, whereupon the machine opens its doors and loudly sings: How dry I am, how dry I am, nobody knows how dry I am. . . .

so connected and interconnected as to allow the simultaneous performance of a trio of cycles by fluid and gaseous refrigerents, endlessly flowing in a closed system of pipes. The basic charge is made up of ammonia, water, and hydrogen, but within the system the ammonia exists in the liquid state, in the gaseous state, and as an aqueous solution which may be either strong or weak. The hydrogen gas is to be found as such, and also admixed with vapourised ammonia.

In the main cycle, the application of heat to an aqueous solution of ammonia contained in the boiler, drives off ammonia vapour. This passes through a water separator (at which point any associated water vapour condenses, and runs back to the boiler) and so reaches the condenser, the cooling fins of which occasion the removal of sufficient heat from the dry vapour to cause it to condense to liquid ammonia, in which state it trickles into the evaporator—the actual cooling unit of the refrigerator, located inside the cabinet. Here, the liquid ammonia (in a manner shortly to be described) is induced to evaporate, which it does with the absorption of considerable heat, thereby lowering the temperature within the refrigerator. The vapour then passes on to the absorber, where it encounters a weak aqueous solution of ammonia, by which it is dissolved. The resultant strong solution of ammonia makes its way to the boiler, *via* a heat-operated bubble pump, thus completing the cycle (Fig. 7—Section A).

In the first of the two subsidiary cycles, a stream of gaseous hydrogen enters the evaporator and there mingles with the trickle of liquid ammonia coming from the condenser, an association which (as the hydrogen makes up the bulk of the pressure) lowers the vapour pressure of the ammonia sufficiently to enable it to evaporate, as earlier recounted. When the mixture of the two gases reaches the absorber, the removal of the dissolvable ammonia gas leaves the hydrogen free to return to the evaporator alone (Fig. 7—Section B). In the second subsidiary cycle, meanwhile, the weak solution of ammonia left in the boiler after the driving out of the ammonia vapour by heat, is gravity fed into the absorber, there to regain its former strength prior to being returned to the boiler (Fig. 7—Section C). As an added refinement, the heating unit (gas or electric) is fitted with a thermostatic device which automatically maintains any desired temperature within the cabinet.

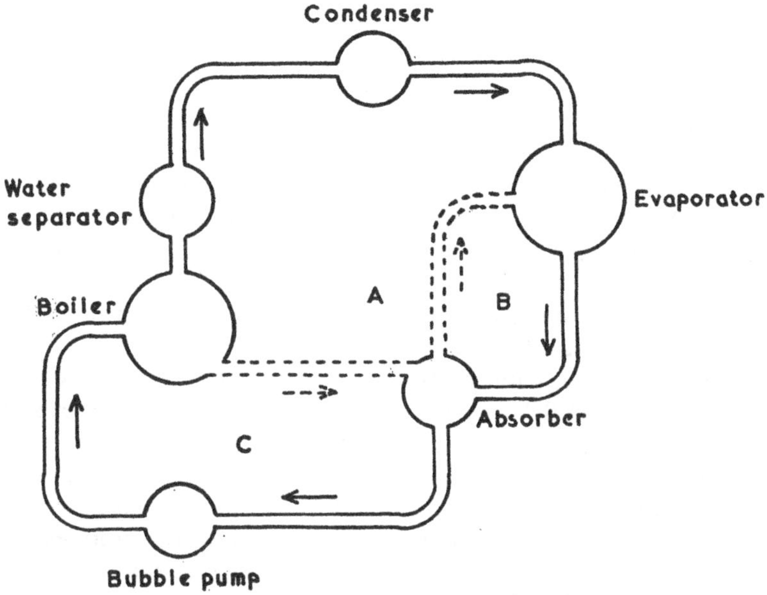

FIG. 7.—Electrolux refrigeration. The main cycle, *A*, is linked with the subsidiary cycles *B* and *C* as shown.

Nor need premises in outlying districts, beyond the reach of mains supplies, be without electricity—or their occupants troubled by a manual stopping and starting of a generator motor, or reduced to the use of storage batteries. The automatic control of Diesel-electric sets devised by Petbow Limited, incorporating a novel 'Start-on-Demand' feature, provides a complete answer to such inconveniences. A touch of a light or power switch anywhere in the premises completes a circuit through a relay, and the generation of current begins forthwith. In this automatic starting-up of the engine, moreover, the controlling mechanism makes due allowance for prevailing conditions, and if the engine happens to be hot or warm from recent use, cranking time is reduced accordingly. Production of current within the capacity of the generator continues as long as one or more switches are in the 'On' position. But as soon as the last switch is turned off, the engine at once comes to a stop, ready to start up again in response to the next operation of a switch.

Electricity supplies apart, modern automatic vending machines are not only considerably more versatile than their cigarette-selling forbears, but they are also much more polite, and some of the latest models say 'Thank you' to their customers, or words to that effect. In the U.S. particularly, these mechanical salesmen offer an extraordinary assortment of services and goods, including a shave (by electric razor), a shoe shine, sunray treatment, and a pre-cooked meal with a wide choice of courses. In return for the appropriate coin, which (men being deceivers ever) the machine methodically and scientifically tests for genuineness before accepting, it is possible to obtain almost anything that is edible, from a fruit pastry to a cup of steaming hot coffee. This last named delicacy is made and offered by machines such as the American National B-600 which, as its numerical designation suggests, has a capacity of 600 cups. It requires no attention other than a periodic replenishment of containers and powdered ingredients, though permanent water and electric supplies are, of course, necessary. The machine is so designed that it will not accept a coin while it is in the process of serving, or when it is sold out. Neither will it dispense free drinks, should money by some mischance lodge in the coin switch!

Another make of machine provides quart-sized paper-board containers filled with refrigerated milk, and though it accepts only one coin, it will give change. Occupiers of blocks of service flats are tending increasingly to patronise these machines when they find one parked near their doorstep, and they are rapidly taking the place of the conventional milk roundsman. The milk in the machine, which is replenished at least once daily, is always fresh, for it is maintained at suitably low temperatures. And not only is it available at any hour of the day or night, but there are no bottles to clean and return. It is also cheaper to buy than the delivered product, and the service is now being extended to other foods which call for refrigeration, such as butter and cream.

In food processing and packing, where the absence of handling is particularly desirable in the interests of cleanliness, countless automatic contrivances are now employed, from the Lynch

CAVALIER *Sunday Dispatch*

L-152 model which makes and wraps soft ice cream sandwiches (at a rate of more than 100 a minute) to the machines made by the Doughnut Corporation of America (and by its English subsidiary), which automatically weigh, cut, cook, float, turn, and deposit doughnuts at the rate of 5,000 or so an hour. Other mechanisms, such as the Cox Z-2, will, by photo-electric means, grade and sort beans and like commodities according to hue and relative brightness, ensuring the removal of any bad or doubtful specimens, together with all foreign matter, from a consignment about to be processed and canned. Similarly, protection against the presence of hidden nails or other metallic impurities in edible products is offered to makers of bread, biscuits, and the like by the Cintel inspection unit, the industrial model of which comprises a search head containing three coils—an oscillator and two pick-ups. These coils are screened from surrounding metal objects (such as other machines) which it is *not* desired to detect, and are so arranged that there is an 'electric balance' between them. The oscillator coil, energised from an amplifier unit, provides an electric field which completely covers a central aperture in the search head, through which opening objects to be searched pass in a continuous procession on a conveyor. Should a piece of metal enter the field set up by the oscillator, the aforementioned state of balance is at once upset, and current is induced in the pick-up coils. This induced current, after amplification, operates a relay, which in turn causes a red light to shine, or a bell to ring, or the conveyor to be brought to a halt. Alternatively, it can be arranged for the offending article to be removed by the machine itself. This is contrived by means of a delay unit which notes the alarm signal, and acts upon it at a point where the procession normally passes over a miniature drawbridge. At the appropriate moment, the alarm signal sets into motion a solenoid-operated arm, which pulls up the bridge and so allows the contaminated article to fall into a reject receptacle. The bridge then returns to its normal position in time to carry the next item.

The phenomenon of induction is also the basis of the Tyer-Integra system of automatic train control, in use on the Swiss Federal Railways. There is no actual contact between that part of the mechanism which is moving, and that which is stationary, and a permanently energised electromagnet, E_1, on the locomotive merely passes over a relaying inductor, U_1, placed on the track (Fig. 8). Current is thereby induced in U_1, and if the signal control relay at S has been operated to complete the circuit, the induced current will energise another inductor, U_2, also located on the track. This, in turn, will induce current in the windings of a second electromagnet on the locomotive, E_2, so bringing into operation the relay R, which controls the train equipment. It will be seen that if the circuit incorporating the signal relay, *S*, is open, there will be no energisation of U_2, and the train controls will be unaffected. By making use of additional inductors, various controls other than a simple 'Stop' or 'Proceed' have been introduced, and the overall arrangement is such, moreover, that the development of a fault in the main wiring will result in a 'Stop' signal, *i.e.*, any defects automatically operate on the side of safety.

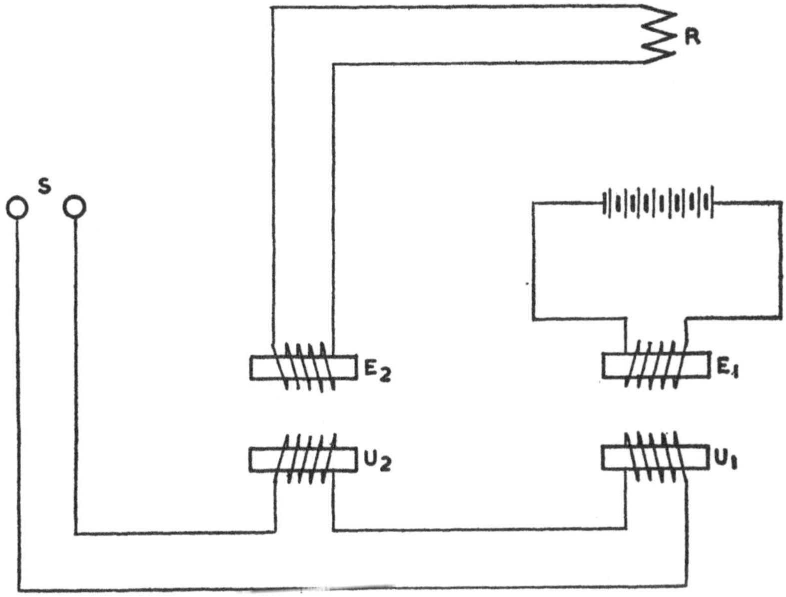

FIG. 8.—Essentials of the Tyer-Integra automatic train control. Note the double transmission of the inductive impulse.

Safety devices on the London Underground are likewise designed to set signals at danger in the event of any breakdown of the signalling mechanism. In one arrangement, applicable where there are no junctions or points, current carried by one of the running rails passes to the other running rail by way of a track relay, so completing a signal control circuit. When a train enters a given section of the line, the current is transmitted by wheels and axles, thus by-passing the relay and breaking the signal circuit. As a result, a signal *behind* the train goes to danger, and so remains until the train moves out of the section concerned.

The question of whether a completely automatic train service could be successfully contrived, in which driverless trains came and went entirely on their own accord, raises interesting possibilities, already anticipated to some extent. The G.P.O. has for years operated a miniature underground mail train in London which has no driver, and as early as 1924, the first year of the Wembley Exhibition, the public was introduced to a passenger-carrying robot railway which boasted the name Never-Stop, and which, true to its title (at any rate during Exhibition hours) at no time came to an actual halt. The carriages were driven by means of an endless Archimedean screw, the pitch of which governed their speed. The arrangement was such that they slowed down to a crawl when passing through stations, and immediately quickened their pace thereafter. The line formed a single loop, and there was, of course, no possiblity whatever of collision.

Automatic signalling devices as applied to road traffic are familiar enough, and the time-wasting propensities of early versions, employing a fixed time cycle, have been effectively overcome by the Electro-matic or vehicular-operated type of control. The customary Red, Amber, and Green light signals remain as before, but are linked electrically with detector units, and with a control mechanism which acts upon signals sent to it by these units. Usually, they take the form of a press-operated pneumatic strip, placed on a level with the surface of the roadway. The weight of a passing vehicle expels air down a tube, and so moves a diaphragm in a capsule located in a junction box under the pavement. This movement closes an electric circuit to the controller.

In this way, a vehicle not only gives warning of its presence, but it also announces the speed of its approach, information which the control mechanism notes and acts upon, having regard for the time of the vehicle's arrival in relation to that of other road users, and the density of traffic then prevailing. Thus, when a solitary vehicle approaches a signal at Green, its contact with the detector has no visible effect. But its passage has nevertheless been observed, and despite any conflicting demands from

PLATE 8(*a*). In this food processing establishment, beans are being automatically graded according to colour. Foreign materials, and all beans which do not exhibit a predetermined standard of whiteness, are promptly rejected by means of a photoelectrically operated sorting mechanism. Photo: *Electric Sorting Machine Co.* (*b*). The Cintel electronic device unerringly detects the presence of metallic impurities in non-metallic substances, such as foodstuffs. It is here shown examining a procession of date blocks. Photo: *Cinema-Television Ltd.*

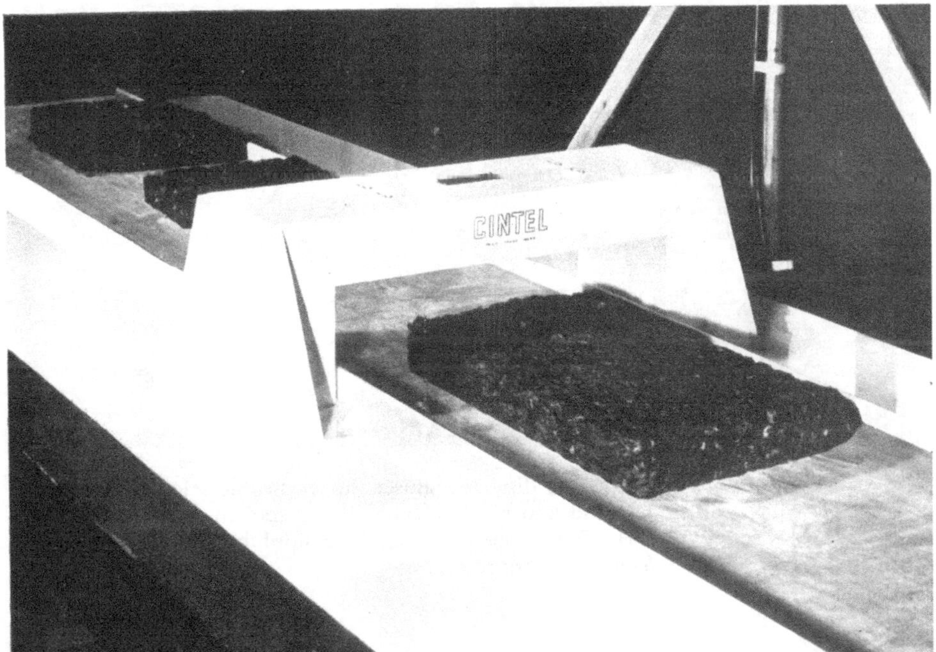

PLATE 9. The automatic digit recogniser, an electronic telephone device which can understand and react intelligently to spoken numbers. It was devised by K. H. Davis (seen above), R. Biddulph, and S. Balashek. Photo: *Bell Telephone Laboratories, Inc.*

other traffic which may have arrived in the meantime, sufficient time for a crossing of the intersection is allotted to the first-comer, the speed of its approach having been taken into account.

What happens when a vehicle approaches a signal at Red depends upon the circumstances of the moment. If there is no other traffic in the vicinity, the Amber light should immediately show, and the signal change to Green in time for the vehicle to proceed with undiminished speed.* If, on the other hand, a continuous stream of traffic is crossing its path, the signals will change in favour of the waiting vehicle immediately a gap in the opposing traffic stream occurs. Should the awaited gap not occur, the signals will arbitrarily bring the moving stream to a halt after a predetermined period of time. And if, meanwhile, a long queue has formed behind the waiting vehicle, this, too, will be permitted to move for a limited period only, though due allowance will be made for the fact that a start has to be made from rest, *i.e.*, a distinction is made between vehicles which are moving, and vehicles which are stationary. But always, the invariable rule is that traffic must surrender a right of way after a maximum interval, if (but not otherwise) there is other traffic requiring it.

A feature of the system is the facility with which it can be adapted to suit special circumstances. Where narrow roads exist, and where vehicles, after crossing an intersection, are liable to run over the detector unit on the far side, the sending of a misleading signal to the controller can be prevented by means of a double contact detector which is uni-directional, and which operates only when it is crossed in a particular direction. Again, along main town or city thoroughfares with many intersections, co-ordinated signals can be introduced, and the principle of repeat-pulsing employed, whereby traffic, on arrival at one inter-

* In fact, this never seems to happen—at any rate to the author. Invariably, the Amber light lingers exasperatingly on, making braking necessary, and a subsequent change of gear inevitable. The explanation is that, so far as the British Isles are concerned, a Ministry of Transport ruling has fixed the duration of the Amber signal at three seconds. As the detector is normally sited some 90 lineal feet from the stop line, it is possible to proceed with undiminished speed only if one is travelling at 20 miles an hour or less.

section, automatically sends advance warning of its impending arrival to the next intersection, so giving notice that a right of way will soon be required.

The brain behind this highly selective process is the controller, which consists, essentially, of a solenoid-operated cam shaft, a number of relays, and two or more timing circuits. The fundamental components of the timing circuits are a gas discharge tube (Neon type), a condenser, and a variable resistance. The apparatus functions as follows: a source of high voltage current is connected, by way of a variable resistance, to a condenser. This charges to the flash voltage of the Neon tube after an interval, the length of which is governed by the value of the variable resistance. A property of the Neon lamp, of which use is here made, is that no current will pass until the voltage across its terminals reaches ionizing point. When this happens, the current thus released actuates one of the relays, and this brings into play the solenoid-operated cam shaft, which in turn controls the signal lights. The situation thus is that a critical voltage is built up over a predetermined period which, on demand, will act in favour of a vehicle confronted by a signal at Red. All the while, however, the condenser charge is being dissipated in part each time a vehicle which *has* the right of way passes over a detector unit, the amount of the charge that is lost depending upon the speed of the vehicle concerned, *i.e.*, upon the duration of the contact that is made by its wheels.

A prolonged lull in this process of charge dissipation would, of course, permit the critical voltage to be reached without interruption, whereupon the demands of any cross traffic would be met forthwith. But if such a lull failed to occur, it will be evident that the flash-over voltage might never be reached, and that an unbroken line of traffic could, by its very continuity, so dominate the situation that a perpetual right of way was established! Such a contingency is guarded against, simply and yet effectively, by the incorporation of a second circuit similar to the first, *i.e.*, there is an identical circuit for each of the two crossroads. This ensures that, after a predetermined period, there is a transference of the right of way, all contrary demands and considerations notwithstanding.

The automatic signalling of motor vehicles is now being followed by the installation of various devices designed to ensure their automatic operation. Among the new transmissions which have been developed in recent years, an outstanding contribution, made by The General Motors Corporation, is the Hydramatic drive, recently adopted, with minor modifications, by Rolls Royce Limited. With its aid, car driving is reduced to depressing the accelerator pedal to go, and applying the brakes to stop.

All this, and more, is accomplished with the aid of three sets of planetary gears, a multiple disc clutch, a pressure-activated governor, a system of hydraulically operated valves, and numerous other devices, the whole forming an intricate assembly far too complicated to permit of a description in detail here. It must suffice to state that the mechanism acts on the principle that for any combination of vehicle speed and accelerator position, there is an optimum gear ratio, and this particular ratio it will automatically select from the four forward speeds at its disposal. As operational flexibility would be unduly restricted if the gear changes always occurred at the same vehicle speed, the arrangement is such that during rapid acceleration, or hill climbing, the transmission lingers in low gear, staying there longer than would normally be the case. On the other hand, there are occasions, *e.g.*, when overtaking in traffic, when it is desirable to achieve a sudden extra burst of speed, and this can readily be obtained by stepping hard on the accelerator pedal until a pronounced resistance is felt, and thereafter depressing it still further. The gears will then automatically change *down*, from fourth to third, or from third to second, as the case may be, so enabling the car to surge forward.

As to future developments, a glimpse of these has already been provided by RCA's Dr. Vladimir Zworykin, whose self-driving car has demonstrated its ability to keep to a circular track, guided by electronic means which will bring it safely to a halt if at any time it threatens to collide with another vehicle ahead. The inventor, famed deviser of the electron camera, foresees the day when such guided vehicles will drive themselves along the highway without anyone at the wheel at all.

Developments no less revolutionary, ranging from automatic broadcasting stations to voice-operated telephones, have in the meantime been taking place in the field of comunication. At Daventry, for instance, B.B.C. transmitters can be switched on or off by remote control. To guard against breakdown, and yet avoid the expense of duplicating all the equipment, the power output is shared between two transmitters, operating in parallel. Should a fault develop in one of them, the other automatically takes over completely until such time as the transmission can be restored to full strength.

In telephony, the automatic system most widely used to-day is based on that devised by A. B. Strowger as long ago as 1889—the step-by-step system. It consists, essentially, of a combination of relays and two-motion selectors. The act of lifting the receiver sets a line switch hunting for a disengaged 'first selector', to which the caller is automatically connected. The subsequent dialling of a number—say 6789—causes a succession of electrical impulses to step up the 'first selector' to the sixth level of a bank of contacts, along which it proceeds to search for an available 'second selector' in this 6000 group. The impulses from the dialling of the next digit step up the 'second selector' to the seventh level of another bank of contacts (700 group), after which a 'final selector' moves up to the eighth row of its bank of contacts (80 group) and along to the ninth contact at that level.

In large cities, the existence of several exchanges inevitably introduces complications, hence the need for additional dialling based on the first three letters of the name of the exchange concerned. And as it is not an economical proposition directly to connect each exchange with every other local exchange, a particular call may have to be routed through several exchanges before the distant subscriber is reached. As may be imagined, the application of the dialling system to trunk calls introduces many more complications still. In America, where Bell Telephone Laboratories set to work on the problem more than twenty years ago, FACD (foreign-area customer dialling) is at last beginning to come into use. The procedure finally evolved, known as the cross-bar system, automatically and successively selects the best available route to the required destination, all the while 'remembering' the number which has been dialled. And details of the call—its duration and its distance—are automatically recorded by mechanisms which compute the cost, and make out a charge slip.

Other recent developments by Bell include a telephone answering set, and a device known as the automatic digit recogniser. The first-named is a robot secretary which can take incoming calls after giving the caller a message. When the instrument is connected to an ordinary telephone, the subscriber can record a message of up to 30 seconds duration, using the telephone itself to make the recording. Thereafter, in his or her absence, the machine will answer a caller with the message, and then switch to a recording condition and note anything that is said in reply, prior to clearing the line (after a predetermined period) in anticipation of the next call. Twenty such messages can be taken during any one period of absence.

The automatic digit recogniser, called Audrey for short, is an electronic contrivance which can understand and react intelligently to spoken numbers. Which is to say, to make a call, the telephone subscriber will in future merely speak the number as he would if he were connected to a manual exchange, the only difference being that the place of the listening operator is taken by the machine, which proceeds to sort out the speech sounds into electrical categories, the electronic equivalent of sound wave patterns. By matching each category in turn against a memory cell containing standard reference patterns, the machine is able to recognise numbers when it hears them spoken.

III

Enter The Computer

MECHANICAL aids to counting, such as the abacus, have been in use for thousands of years. But the first real calculating machine was not invented until the middle of the 17th century, when Blaise Pascal demonstrated his contrivance for performing addition and subtraction by means of wheels bearing the numbers 0—9, so arranged that as one wheel moved from 9—0, that immediately to the left of it moved through one tenth of a revolution. Morland followed with a similar device, and so did Leibnitz, who subsequently produced a multiplier with a stepped wheel, operating on the now familiar basis of repeated additions.

A more ambitious plan, conceived by Charles Babbage in 1812, concerned the construction of a 'difference engine'. The machine was intended to facilitate the determination of tabular numbers by a summation of successive differences:

1		8		27		64		125		216		343. . . .
	7		19		37		61		91		127. . . .	
		12		18		24		30		36. . . .		
			6		6		6		6. . . .			

It will be observed that when each of the numbers 1, 8, 27, 64, 125 . . . (*i.e.*, the cubes of the integers 1, 2, 3, 4, 5 . . .) is successively subtracted from the one which follows it, the result is 7, 19, 37, 61 . . . (first order of differences), and that a repetition of the process gives 12, 18, 24 . . . (second order of differences), which in turn yield a constant (6). Conversely, tabular numbers such as the cubes aforementioned may be obtained by a series of simple additions: $6+12=18$; $12+7=19$; and $7+1=8$, *i.e.*, 2^3; $6+18=24$; $18+19=37$; and $19+8=27$, *i.e.*, 3^3; $6+24=30$; $24+37=61$; and $37+27=64$, *i.e.*, 4^3; and so on.

The machine was designed automatically to undertake these additions, and work on the device, which had the backing of the British Government, was begun in 1823. But Babbage revised his plans with such frequency that the project was still incomplete after the passing of ten years and the spending of some £17,000, at which juncture a halt was called. The enterprise was finally abandoned in 1842.

The first successful calculator to be produced commercially, the Arithmometer of C. X. Colmar, had in the meantime made its appearance, and half a century later came W. T. Odhner's machine, which provided the basic design for a host of mechanisms since marketed in many different lands under a wide variety of names—Brunsviga, Rapide, Facit, Calco, Marchant, Britannic *et al.*

Such desk machines, even when electrically operated, are not, of course, to be compared with more elaborate systems such as the Hollerith,* in which a punched card takes the place of the book entry in manual accounting. The punched holes, by virtue of their position on the card, represent various items of information—a number, a date, a sum of money. Preparation of the cards may be carried out by automatic punches, the keyboard of which resembles a simplified typewriter with twelve main keys, marked 0—9, and 10d and 11d respectively. As the various keys are depressed, the action is sensed electrically, and information is conveyed to a storage unit where it is retained pending further instructions from the operator. The machine, in other words, is able to memorise data, thus giving the operator an opportunity to amend conscious errors (by depressing a cancellation key) before the card is actually punched. Thereafter, at the touch of a punch bar, perforation proceeds in accordance with the instructions already given, and the card is automatically ejected when the operation is complete, a new card taking its place. The work of recording, meanwhile, goes steadily on.

The next step is to eliminate any errors which may have been

* 'Hollerith' is the registered trade mark of The British Tabulating Machine Company Limited, which was formed in 1907, and its American counterpart is the International Business Machines Corporation, represented in this country by IBM United Kingdom Limited.

made by the operator unwittingly, and this is done with the aid of a companion machine known as the Verifier. The punched cards are self-fed into this machine, and a second operator goes through the motion of re-punching the cards, reading from the original source of information. The Verifier, however, does not actually punch holes—it merely compares the existing pattern of perforations with that indicated by its own key depressions. Should these not match, the machine immediately stops, and an error signal light glows red. Cards which are passed as correct are notched with a small 'v' on the top left hand side, and then automatically ejected.

After verification, the cards, in random order, are hopper fed into a sorting machine, through which they pass in a continuous stream, a metal brush completing an electrical circuit each time it senses a perforation. The machine does the rest, automatically separating the cards in accordance with the instructions thus transmitted to it, and delivering them to one of thirteen pockets. Of these, twelve represent significant positions in the card columns, and the thirteenth is for rejects. In this way, it is possible, in the space of one hour, to deal with 24,000 cards—and if this is not quick enough, a High-Speed Sorter, utilising an electronic device, will do the job even faster.

The cards, thus sorted, form a permanent basic record, available at any time for transference to the Tabulator, the machine which extracts the punched information they contain, and presents it for consideration in readable form. This mechanism, merely by sensing the perforations in the cards presented to it, selects data relevant to desired results, memorises figures, adds and subtracts, compares its findings at pre-determined stages of the operation, and presents final details (totals, balances, and grand totals) in printed words and figures. The machine's nerve centre is the control panel, in effect a telephone switchboard in miniature. By inserting plugs in the panel, connections appropriate to the operations required can readily be made, and the electrical impulses picked up from the cards thus directed to the various devices within—the adding counters, the storage units, and the printing banks—so that the necessary functions automatically take place.

Nor is this all. Supplementing the machines just described are many others*—the Gang Summary Punch which, among other tasks, prepares summary cards simultaneously with tabulating operations; the Reproducing Punch, which reproduces punching from one set of cards to another; the Mark Sensing Reproducing Punch, which converts data pencilled on to cards into punched holes; the Balancer, which provides preliminary totals for balancing purposes; the Collator, which automatically associates two sets of cards in one operation, and likewise segregates one pack of cards into two (not on one column, but on the number of columns occupied by the controlling data); the Interpreter, which translates punched holes into alphabetical or numerical characters, at the same time printing the results along the top of the card; the Card Counting Sorter, which counts the perforations in any or in all positions of a given column, and at the same time arranges the cards in numerical sequence; the Type 504 Multiplier (a combination of relay and mechanically driven counter wheels), which automatically multiplies two numbers punched in a card, and punches the product into the same card; the Electronic Multiplier, which performs similar functions, and has an output of 6,000 cards an hour; the Electronic Calculator, which is capable of performing a remarkable variety of all four arithmetical functions, completing and checking an entire series of calculations in any desired sequence, and punching the product in card form; and the Hec Electronic Computer which, by the magic of its performance, surpasses them all.

Electronic computers are the outcome of substituting the thermionic valve for the relay, long used to control complicated switching procedures employed in contrivances endowed with an ability to count, to translate, to locate, to select, and to calculate. But whereas electro-mechanical methods incorporating the relay enabled additions to be performed at a rate of 500 or so a minute, the introduction of the thermionic valve at one step increased this figure to no less than 300,000 a minute!

* There is even a machine for converting punched cards into perforated tape, and *vice versa*, so that recorded data can be transmitted by G.P.O. lines from isolated localities to centralised points.

This tremendous advance took place in the 1940's, when ENIAC (Electrical Numerical Integrator and Computer) was planned (1942) and completed (1946) at the Moore School of Electrical Engineering at the University of Pennsylvania, in the U.S. It used more than 18,000 valves, was programmed by means of plugs and switches, could multiply two ten-figure decimal numbers in less than 1,000th of a second—and is already obsolescent!* Its successors, EDVAC (Electronic Discrete Variable Automatic Computer), BINAC (Binary Automatic Computer), and UNIVAC (Universal Automatic Computer), not to mention the British ACE (Automatic Computing Engine), and an ever-increasing number of smaller machines intended for commercial use (of which the Hollerith Hec model is an example), have abandoned the decimal system in favour of the binary or dyadic, the base of which is 2.

The use of 10 as a base is undoubtedly a survival of an ancestral habit of counting on fingers, and the Creator, had He been so inclined, could have bestowed considerable benefit on mankind by endowing each human hand with six terminal parts instead of five. As it is, we find ourselves with a base which suffers from the disadvantage that it has only three exact divisors, whereas 12 would have given us no less than five—1, 2, 3, 4, and 6. Proposals to remedy the consequences of this omission, by introducing two new digit symbols and so establishing a duodecimal system, have been made from time to time. But, human nature being what it is, the decimal habit is with us yet.

Although the use of binary numbers in electronic computation is a comparatively recent innovation, such a dual notation, employing only the digits 0 and 1, is actually more than 5,000 years old, as Leibnitz discovered when he deciphered the ancient

* ENIAC continues in operation at the Ballistic Research Laboratories at Aberdeen Proving Ground, Maryland, and it has been modified considerably since its installation there in 1947, the latest improvement being an extensive addition to the memory system. It was early used to calculate the values of the transcendental numbers *pi* and *e* to two thousand decimal places, and more recently it has solved many of the trajectory problems of the new 280 mm. atomic cannon—each in less time than is required by the shell to travel from gun to target!

taken from
.*e*., 1, 2, 4, 8,
,234 means
-4), so the
1, One, One,

$0.2^2+1.2^1+$

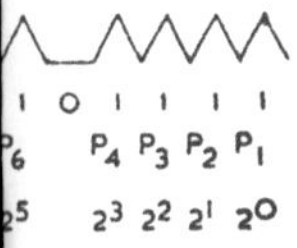

are repre-
e decimal
$^1+2^2+2^3$
$^{25}+2^{30}+$
1.

comprise the
o the every-
e inordinate
ld very soon
, One, One)
99! Such a
o a machine

the rule for
e divisor must

raised to the

ombinations), and so
tained. Compare:

Yi-King

It will be observed that successive binary digits right to left, are powers of 2—$2^0, 2^1, 2^2, 2^3, 2^4, 2^5$... (16, 32...).* And just as the decimal number $1.10^3 + 2.10^2 + 3.10^1 + 4.10^0$ (*i.e.*, $1{,}000 + 200 + 30$ binary number 10,011,010,010 (read, 'One, Oh, O Oh, One, Oh, Oh, One, Oh') may be expressed as:

$$
\begin{aligned}
&10{,}011{,}010{,}010 \text{ (binary)}\\
&= 1.2^{10} + 0.2^9 + 0.2^8 + 1.2^7 + 1.2^6 + 0.2^5 + 1.2^4 + 0.2^3 +\\
&\quad 0.2^0\\
&= 1{,}024 + 128 + 64 + 16 + 2\\
&= 1{,}234 \text{ (decimal).}
\end{aligned}
$$

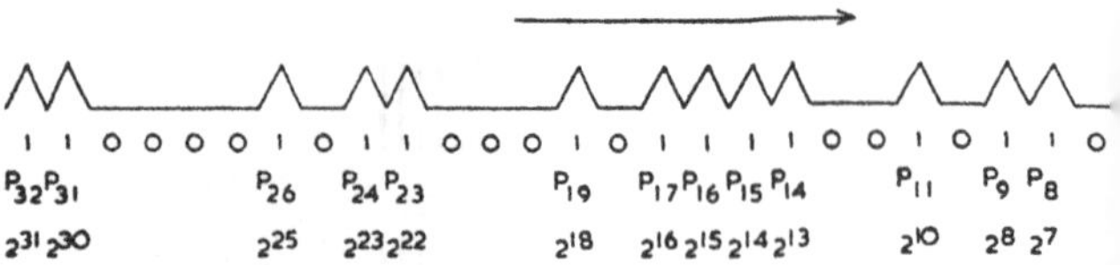

FIG. 9.—A typical pulse train, in which 1 and 0 sented by the presence and absence of a pulse. Th equivalent of this binary number is the sum of $2^0 +$ $+ 2^5 + 2^7 + 2^8 + 2^{10} + 2^{13} + 2^{14} + 2^{15} + 2^{16} + 2^{18} + 2^{22} + 2^{23} +$ $2^{31} = 3{,}267{,}749{,}295$. Note that P_{32} represents 2^{31}, *i.e.*, 2^{32}

Binary multiplication and addition tables merely rules $1 \times 1 = 1$, and $1 + 1 = 10$. An evident obstacle day use of so delightfully simple a notation is th length of the expressions it would involve—we shou tire of repeating 1,100,011 (One, One, Oh, Oh, O every time our medical adviser wanted us to say length of expression, however, offers no difficulties

* By mathematical convention, $2^0 = 1$. This follows fro the division of powers of numbers, that the exponent of th be subtracted from the exponent of the dividend. Thus:

$$X^n \div X^n = X^{n-n} = X^0 = 1.$$

Since X may be any number, it follows that any numbe zero power is 1.

operating at the speed of light, while the use of only two digits greatly simplifies a translation into momentary electrical charges or pulse trains, in which 1 and 0 are represented by the presence and absence of a pulse (Fig. 9). Moreover, it can easily be arranged for the machine to convert decimal numbers into binary, and then back again into decimal, automatically and entirely on its own account.

The essentials of the electronic computer are the aforementioned pulse trains (generated at the rate of a million a

KIRKBRIDE *Sunday Dispatch*

This is the latest thing in adding machines!

second); a memory system (for the storage of information and instructions); a control (for supervising the carrying out of those instructions); valve circuits (so arranged as to be able to combine pulse trains in accordance with arithmetical laws) (Fig. 10); and input and output devices (by means of which the mechanism receives its instructions and announces its findings).

At the start, numbers relating to a given problem, and instructions in regard to them, are fed into the machine in codified form, *i.e.*, the machine is programmed—by means, say, of punched cards or perforated tape. Trains of electrical charges are thereby set up, and are distributed to, and stored up in, the memory circuits, whereupon work on the problem begins. The first instruction is automatically transmitted from the storage unit to the control, and is at once carried out by the arithmetic section, after which the next instruction is issued and executed, and so on, until the entire operation is completed. In computer terminology, each machine has what is known as an

ORDER CODE, or basic list of instructions, and this, together with data relating to a given problem, makes up the

PROGRAMME, which is fed into the machine's memory system, comprising a great many

CELLS, each one of which has a specific

ADDRESS, and is capable of accommodating one

WORD, or pulse train, which may represent either an instruction or a number.

In as much as both numbers and instructions are presented numerically, the machine is not able to distinguish between them. But whereas a number treated as an instruction would have disastrous consequences, part of the art of programme devising lies in arranging for certain instructions to be treated as numbers, whereby new instructions are produced. In effect, the machine can be induced to amend its programme as it goes along, thereby achieving mathematical short cuts!

Storage of numbers and instructions is accomplished in several ways—for example, by the use of supersonic mercury delay lines. Here, a given pulse train is converted into compara-

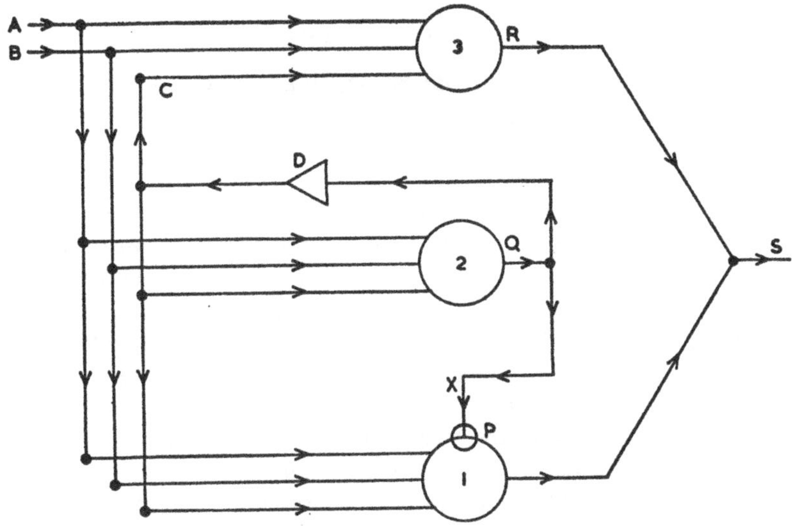

FIG. 10.—An adding circuit. The four components, R, D, Q, and P represent assemblies of valves and other electronic devices, so arranged that a given pulse, P_i, will emerge on the componental output line if:

(*a*) in the case of R, P_i occurs on all three input lines, A, B, and C;

(*b*) in the case of D, P_{i-1} occurred on the input line;

(*c*) in the case of Q, P_i occurs on any two or on all three input lines, A, B, and C; and

(*d*) in the case of P, if P_i occurs on one or more of the three input lines, A, B, and C, but does *not* occur on the inhibiting line, X.

The four components are so connected that if two numbers enter on lines A and B in the same minor cycle, their sum will emerge on line S. Thus, if in the time interval i, $A=1$, $B=1$, and $C=0$, then:

(*e*) R receives P_i on 2 lines, and so does not emit a P_i;

(*f*) Q receives P_i on 2 lines, and emits P_i to D and P; and

(*g*) P receives P_i from Q on the inhibiting line, and does not emit a P_i.

As a result, S does not carry a P_i, and at time $i+1$, D will emit a P_{i+1}, together representing $1+1+0=10$. (With acknowledgements to R. L. Michaelson's Paper, *Large-Scale Electronic Digital Computing Machines*, published in *The Journal of the Institute of Actuaries*, Vol. LXXIX, Part III, No. 353, December, 1953).

tively slow moving sound waves, which are made to travel from one end of a mercury filled tube to the other. They are then reconverted into a pulse train, amplified, and promptly turned back into sound waves again—a cyclic process which can be continued indefinitely. In another system, the pulse train is distributed as a charge pattern upon the surface of a cathode ray tube, from which it may be subsequently retrieved. And a third method involves the use of a magnetic drum or tape, and the representation of digits by different degrees of magnetisation which, once set up, will last indefinitely, an advantage not shared by other types of memory.

But precisely how are these pulse trains made to simulate numbers? The heart of the electronic computer is its valve system, and this may be regarded simply as an assembly of electronically operated switches. In partnership, two such valves can be arranged to represent a switch which is either Off, or On (flip-flop circuit). When the device is in the Off position, the arrival of an electrical pulse will change it to On. And when it is in the On state, the next pulse will change it back to Off *and a pulse will be emitted in the process.* Suppose, in order to make use of the emitted pulse, we link three of the valve couples in tandem. And suppose, furthermore, that we assign the symbol 0 to the Off position, and the symbol 1 to the On position. In response to a continuous series of pulses, the initial and subsequent situations (reading from right to left) will be:

0 0 0	initial state
0 0 1	first pulse received
0 1 0	second pulse received
0 1 1	third pulse received
1 0 0	fourth pulse received
1 0 1	fifth pulse received
1 1 0	sixth pulse received
1 1 1	seventh pulse received
0 0 0	eighth pulse received

But here, it will at once be apparent, we have a representation in binary form of the decimal numbers 0—7! It will be evident,

PLATE 10(*a*). ENIAC—general view. Photo: *U.S. Army*. (*b*). UNIVAC. Like its predecessor BINAC, it comprises two identical machines, and can check its calculations as it goes along. Its memory can store 12,000 digits or alphabetical characters in 1,000 locations, and each location is available to the computing circuits in a time of 40–400 microseconds. Photo: *Remington Rand, Inc.*

PLATE 11. Rear view of ENIAC, showing some of its thousands of valves. Note the ventilating ducts and the portable rack containing valve replacements. Photo: *U.S. Army*.

moreover, that decimal numbers higher than 7 can readily be accommodated by adding to the line of valve couples. If their number were increased from three to four, decimal numbers up to 15 could be represented (1,111 binary); if from four to five, up to 31 decimal (11,111 binary); if from five to six, up to 63 decimal (111,111 binary); and so on.

The digital position (or power of 2) is so related to the time scale that the presence of a pulse in the *n*th interval represents 2^{n-1}, and the time required for the transmission of a pulse train, $P_1, P_2, P_3 \ldots P_n$ is termed a minor cycle. The Hollerith Electronic Computer (the existing model of which employs 550 valves) has no less than 32 couples in line, so that a standard pulse train can represent any binary number up to 11,111,111,111,111,111,111, 111,111,111,111, *i.e.*, up to $2^{32} - 1$, or 4,294,967,295 decimal. And if this does not suffice, arrangements can be made to handle even larger numbers.

The ACE machine is similarly endowed, though its total valve complement is about 1,000. The existing machine is a pilot model, the construction of which, under the administrative supervision of F. M. Colebrook, was begun in 1947 by a team of four mathematicians and ten electronic engineers of the National Physical Laboratory, assisted by a small group of production engineers from the English Electric Company. The second ACE (known also as the Deuce), which will be larger and faster than its prototype, is still in the course of development at the time of this writing. It is planned to add to the existing storage capacity of 8,000 binary digits (held in supersonic delay lines) by means of an auxiliary magnetic drum.

Other British calculators include the EDSAC (Electronic Delay Storage Automatic Calculator) at Cambridge University, and MUC (Manchester University Computer) installed at Manchester University by Ferranti Limited. Machines similar in design are now being produced by this Company for general use. Programming is by means of punched teleprinter tape, feeding up to 200 characters a second, and there is a dual memory system, capable of storing 10,000 binary digits on cathode ray tubes, and 650,000 more on magnetic drums. Addition and multiplication times are 1·2 and 2·2 milliseconds respectively.

In the U.S., the International Business Machines Corporation, which sponsored the SSEC (Selective Sequence Electronic Calculator), and achieved commercial success with its smaller Type 604 Electronic Computer, has now produced a successor to SSEC, intended for industrial use, designated Type 701. It incorporates three different kinds of memory unit (magnetic drum, magnetic tape, and cathode ray tube), and the customary automatic conversion from decimal–binary, and binary–decimal, is employed. The machine is programmed by means of punched cards, and performs multiplication and division at the rate of 2,000 a second, and addition and subtraction eight times as fast.

Another concern which produces a wide range of machines is the Computer Research Corporation of California. The Corporation's Model CRC 102, located at the Massachusetts Institute of Technology, has now given rise to the CRC 102–A, a similar machine which is available for outright purchase (many firms prefer to lease their machines rather than sell them) at a cost of some £22,000. Programming is by punched tape, or the machine can be hooked up to a summary punch using IBM cards. The main memory is of the magnetic drum type, but up to eight CRC 126 Magnetic-Tape units may be connected if it is desired to supplement the capacity of the drum.

The indications are that the electronic computer is fast becoming an indispensable aid to industry and research. The value of the device is not that it can solve problems beyond human capabilities, but that it can produce the answers in as many minutes as would require months, or even years, of work with the aid of pencil and paper alone. F. M. Colebrook, of the ACE team, quotes as a typical case the mathematical complexities which are involved in the designing of airplane wings. The problem might ultimately reduce to the finding of a solution of 50 sets of simultaneous equations, each with the same 50 unknowns. No less than 40,000 separate multiplications would be called for, and a mathematician, working a seven hour day, and even if equipped with a desk calculator, could not hope to complete the task in under six months of continuous computation—as against ten *minutes* for the ACE!

The saving in time and labour is thus enormous. Indeed, tasks are now being undertaken which hitherto went unattempted, simply because of the years of unremitting drudgery which would have been entailed. But are the results obtained to be accepted as correct? In other words, do electronic computers ever make mistakes? The answer is that, on occasion, they do. But all computation is liable to error, and the time-honoured process of cross-checking can be carried out electronically with a facility equal to that which attended the original calculation.*

In the meantime, improvements in design and methods continue to be made. Just as the thermionic valve ousted the relay, so the indications now are that the valve may ultimately be replaced by the transistor. ENIAC's huge and expensive valve assembly not only consumes much power and occupies a great deal of space, but it generates a lot of unwanted heat, the dissipation of which requires an elaborate system of air ducts and motor driven fans. Electronic computers already make extensive use of germanium diodes, and of the many concerns now experimenting with electronically active materials, the Radio Corporation of America has recently announced the construction of transistorised computer elements such as counters and adders, each of which shows great saving in weight, size, and power consumption. Compare:

	Electronic counter using	
	Transistors	**Thermionic valves**
Power consumption (watts)	2	60
Weight (pounds)	½	4
Size (cubic inches)	20	200

Similarly, the experimental RCA electronic adder, capable of 100,000 single-digit additions a second, requires one-quarter the power and occupies one-seventh the space of a comparable computer using the conventional valves. The revolutionary nature of the trend here foreshadowed will be evident.

* Thus the aforementioned BINAC, constructed in 1947 by the Eckert-Mauchly Computer Corporation (now a subsidiary of Remington Rand Incorporated) comprises two identical machines, one of which checks the other. BINAC's successor, UNIVAC, is also a dual mechanism, able to compare each step in its calculations as the operation proceeds.

IV

From Matchlock to Deadlock

AMONG weapons of war, as among machines generally, automatism may be regarded as mechanisation carried to its logical conclusion. From such primitive arms as the club, the spear, and the bow and arrow (an assortment which goes back to Neolithic times and beyond) there emerged the catapult, the trebuchet, and the ballista, outsize hurling devices whose ammunition ranged from rocks and darts to incendiary compositions and the bodies of defunct animals. At this juncture, gunpowder began to be used for military purposes, and thereafter the path of progress is signposted by such names as matchlock, wheelock, flintlock, breech-loader, and repeating rifle,* not to mention the equally significant designations revolver and automatic (though the so-called automatic pistol is usually self-operative only to the extent of being self-loading).

The first fully automatic firearm was the Maxim, named after its inventor, and developed in the 1880's. Earlier machine-guns, such as that devised by Gatling, were in part hand operated. In the Maxim gun, however, the recoil from the discharge was ingeniously utilised to eject the spent cartridge, reload, and fire in a series of actions which ensured continuity of operation by mechanical means alone. The rate of fire was an unprecedented 500 rounds a minute, and such was the amount of heat generated that the barrel of the gun had to be water-cooled.

The principle of self-loading and firing was in due course extended to heavier guns, and in recent years the advent of radar (whereby the distance and direction of a reflecting object can be determined with precision by timing the echo of radio waves)

* In this, as in many other fields, the genius of Leonardo da Vinci (1452–1519) long anticipated actual development. He envisaged breech-loading cannon and rifled firearms centuries before they came into use.

and the electronic computer has enabled the functions of spotting and aiming to be undertaken by the machine as well. In the American 75 mm. A.A. Skysweeper, gun, radar, and computer have been combined on one mounting for the first time. The radar component, operating over a distance of 24,000 yards (15 miles), promptly locks on to an approaching target while the computer proceeds to predict its future position. The gun fires automatically in the light of these determinations, and is claimed to be effective, day or night, against aircraft travelling at speeds of up to 900 m.p.h. at a range of between 360 and 7,200 yards.

The control of interceptor planes is also being increasingly delegated to automatic mechanisms which can react, more quickly than any human pilot, to the changing circumstances of the moment, a necessary development when the high speed of modern aircraft is taken into account—the (piloted) Douglas X-3 research plane was designed to fly at 2,000 m.p.h. In the new supersonic jet fighters, automatic instruments within the plane will take over at the turn of a switch, and relieve the pilot of any obligation to track down his target, or even to aim and fire his weapons. And as to these, the machine-gun and the cannon are now being ousted by air-to-air rocket projectiles such as the Douglas Sparrow (which has a range of 5 miles, and a speed reportedly three times that of sound), and the Ryan Firebird (radar controlled, and fitted with an electronic device designed to guide the missile to its target).

The next and obvious step of eliminating the pilot altogether has already been taken in such remote-controlled machines as the Fairchild Lark, the Convair Terrier, the Boeing Gapa, and the Douglas Nike. This last-named is a radar-operated anti-aircraft missile which forms an integral part of a complex electronic guidance system, able to pick up and keep track of an approaching plane long before it becomes visible to the human eye. All the while, account is taken of the changing position of the intruder, and as soon as the rocket's deadline has been crossed, the waiting missile is automatically despatched, to reach the target area in a matter of seconds.

There are a number of ways in which such a projectile can be induced to seek out its prey, once it has come within a certain distance of it. Homing devices have been developed which are stimulated by the presence of the victim-to-be—by the sound (acoustic waves) of its passage through the air, or by the heat (infra-red rays) emanating from its propulsion unit—and immediately the distance between pursuer and pursued has been sufficiently reduced, other self-operating devices within the rocket (such as a radio fuse) ensure the detonation of the missile's explosive charge. Complications arising from the possibility that the rocket might fail to come within effective reach of its target, or might even attempt to 'home' on to a friendly aircraft, can be countered by various contrivances designed to ensure the safe disposal of such an aimless wanderer.

Eventually, batteries of these guided missiles may well replace heavy anti-aircraft guns entirely.* In the U.S., Nike installations have already been incorporated in the country's defence system, either as permanently stationed groups, or in the form of mobile units, readily transportable by air. And elsewhere, the British Minister of Supply has recently revealed some details of Great Britain's contribution to these developments, giving news of a 2,000 m.p.h. interceptor rocket, booster launched and electronically controlled, and fitted with a homing device which no piloted airplane could hope to outmanoeuvre. The implication here contained was plain enough, and the Press promptly hailed the weapon as the long-sought for answer to the atom bomber—unmindful of the fact that in the days ahead, the atom bomb is not likely to be delivered by piloted aircraft at all.

Just as the tendency in defence is to substitute robot ground-to-air missiles for the interceptor plane and the A.A. gun, so the new weapons of aerial attack are being freed from the limitations which the presence of human cargo imposes. The German V-1, the so-called flying bomb, was in effect a miniature, jet-propelled aircraft carrying a high explosive freight, the whole constituting an expendable unit. But although improved versions of this ingenious mechanism have since been produced, *e.g.*, the Glen Martin Matador B-61 pilotless bomber (reportedly able to

* Since these words were written, the virtual abolition of Anti-aircraft Command has been announced by the British Minister of Defence.

attain 700 m.p.h., and to have a range of 500 miles), such machines would be no match for the more agile missiles which would be sent to intercept them, unless, of course, they could be despatched

TEAL *Sunday Dispatch*

Cissy!

in such vast numbers as to swamp the defence. It would thus appear that the ultimate successor to both piloted and pilotless bombing planes will be the inter-Continental rocket, still in the course of development (*vide* the Atlas project, sponsored by the U.S. Air Force). Once it appears, in the shape of a self-propelled missile with a maximum range of 12,500 miles, *i.e.*, half the equatorial circumference of the earth, no town or city that is not located deep underground will be safe from an ever-present threat of long-distance bombardment.

Exactly how great a menace is this threat likely to prove? In the course of the V-2 attack on Southern England (September 1944–March 1945), some 2,000 rockets were launched, of which about half reached this country. Of these, some exploded harmlessly in open fields, while others claimed as many as 150 victims or more. All told, nearly 3,000 people were killed—considerably fewer than the number of unsuspecting Londoners who were silently gassed to death during four *days* in December, 1952, when as a result of exceptional atmospheric conditions, fumes from power house, factory, and domestic chimneys reached a concentration somewhat more poisonous than usual.

Indeed, if the victims of machines in the present century be classed either as exterminations deliberately contrived, or fatalities accidentally caused, it will at once be seen that the dangers of war have been far outweighed by the perils of peace. In the U.S. during 1953, more than 38,000 persons were killed in road accidents alone, and nearly 1,500,000 others injured—a slaughter considerably in excess of that sustained by all the American Armed Forces in Korea (13,583 missing, 22,731 killed, 105,961 wounded). Again, throughout the war years 1939–1945, industrial casualties in Great Britain were at the rate of 22,000 a month, more than twice as high as those suffered by the fighting Services.

But although carelessness in the handling of machines has, in recent years, resulted in far more deaths and disablements than all man's outbursts of premeditated homicide, the position may soon be changed. The reappearance of the rocket as a weapon of war in modern guise, and its coming use as a carrier for thermonuclear devices, provide a fearsome combination which promises to dominate the entire military situation from now on.* Henceforth, indeed, the maintenance of vast armed forces will be as unnecessary militarily as it is undesirable socially—not only will there be no fighting for such forces to do, but they will not even be called upon to undertake what would be a pointless and hazardous occupation of conquered territory in the shape of a lifeless stretch of radio-active desert. Thus, in the years ahead, it may be expected that the armies, navies, and air-forces of the world will be disbanded as useless against the threat of robot mechanisms which an aggressor, aided by a relatively small band of technicians, could without warning aim at the chosen targets presented by one or more intended victims. There would be no question of successfully warding off such a blow, once the attack had been launched, and only the knowledge that despite the inconceivable devastation it would cause, retaliation might still be possible, could be expected to stay the belligerent's hand. The ultimate situation may well be such that the approach of an air-borne fleet of hostile rockets will be the signal for the radar-operated discharge of a long-prepared counter-blow, even before the impending attack can develop, and we are thus presented with a picture of two opposing camps, each in a position to encompass the total destruction of the other by the mere pressing of a few control buttons, and each fearful lest any untoward move precipitate the downfall of them both.

Is such a stalemate possible? Which is to say, is the combination of rocket and hydrogen bomb likely to prove as destructive as it has been made out to be? In the past, there has always been a tendency to exaggerate the lethal propensities of a new and untried weapon, and to greet its arrival with protests that the race faces annihilation, and demands that the innovation be outlawed in the interests of human survival. And invariably, the supposed precursor of universal doom has turned out to be

* The conventional powder rocket has a long history as a war weapon, and as late as 1806 an improved version of it, devised by William Congreve, was used by British Naval forces to bombard Boulogne. An account of the research which led to the development of the German V-2 liquid fuel rocket will be found in the author's *Into Space* (George Allen & Unwin, Ltd., London, 1953).

nothing of the sort. In a papal bull issued in 1139, even the cross-bow was denounced as a weapon too barbarous for one Christian to use against another, albeit the infidel was exempt from this charitable dispensation, as Brigadier O. F. G. Hogg has remarked.*

Are we to believe, then, that in nuclear fission and its associated phenomena, an instrument of wholesale extermination has at last been found? The answer to this question would appear to be in the affirmative. It is not merely that the setting off of a hydrogen bomb is equivalent to the detonating of millions of tons of T.N.T., or even that the explosion is accompanied by death-dealing radiation. Tremendously destructive of life and property though such effects would undoubtedly be, they would nevertheless be confined to the target areas, so that even in a conflict on a global scale, large portions of the earth's surface would emerge unscarred. The real threat to mankind is to be found in another and more enduring characteristic of nuclear explosions—the creation of synthetic by-products such as carbon 14, a radio-active isotope with a half-life (*vide infra*) of more than 5,000 years. Every time an atomic explosion occurs, the atmosphere is contaminated by such radio-active dust, *and the effect is inescapably cumulative.* Sooner or later, even experimental peace-time explosions will bring the general level of terrestrial radioactivity up to a dangerously high point, while the reckless abandon of an all-out atomic war would soon render the earth uninhabitable.

What follows has been written on the assumption, however unfounded, that mankind will somehow contrive to avoid such a fate—if only to meet with a no less ignominious end, as will in due course be recounted.

* In his contribution, 'Engines of War', to *Chambers's Encyclopedia* (George Newnes Ltd., London, 1950). An echo of this pious disregard for the welfare of the unbeliever was provided by the machine-gun patented by James Puckle in 1718, in which provision was made for the firing of round bullets at the Faithful, and square ones at the Turks!

PART THREE

Machines Minus Men

★

I

Autofabrication

NUMEROUS automatic machines are already at work which constitute miniature manufactories in themselves—the I.R.C. continuous viscose rayon spinning machine (earlier described); the all-in-one Wunderlich spring making machine (which coils, crimps, knots, and heat treats wire in a single operation); the massive Bartref Press (which methodically converts resinated wood waste into compact building board, cut to length from a continuously formed strip); and the Schmutz Four Colour Automatic Cut, Print, and Fold Machine (which manufactures cotton or jute bags at a rate which varies, according to size, from 3,000 to nearly 60,000 an hour)—and the logical outcome of such multi-operational mechanisms is the fully automatic factory, producing a never-ending flow of finished goods from a continuous supply of raw materials.

Excluding such manufactural trivia as shoe horns, knitting needles, letter openers, ash trays, spatulas, toothpicks, and the like (all of which can be moulded from, or stamped out of, a single material in one operation, and thus readily lend themselves to production *en masse*), the construction of complexities such as radios and automobiles solely by automatic means appears likely to be achieved in at least two distinct and successive stages—the making and assembling of parts, and the preparation of materials for component manufacture.

It will be observed that the order of this listing is from finish to start, an inversion dictated by the practical consideration of relative ease of attainment. To contemplate a localised extraction of metals from their ores at the present time, for example, even supposing that all the processes involved could be made self-operational, would be open to a variety of objections, and might well lead to such extravagant conceptions as an automatic nib stamping machine with a steelworks and a rolling mill attached. On the other hand, the supplying of prefabricated components to a production line concerned merely with the task of assembly is an obvious development, comparatively easy to arrange, and one which has already been widely adopted. In the realm of packaging, for example, there are numerous self-operating mechanisms which will prepare a weighed portion of this or that commodity, and then add a wrapper, or make a cardboard carton before proceeding to fill and seal it (Fig. 11). And where existing containers (such as bottles) are to be re-used, there are a host of ancillary devices designed to attend to the preliminaries of cleaning, sterilising, and drying prior to filling, and thereafter to undertake such tasks as labelling, sorting, and crating. How these various operations may be combined, so as to form a continuous production line, is well illustrated by the New Bottling Stores of Ind Coope and Allsopp, at Burton-on-Trent, designed to cope with an unending stream of up to a million empty bottles a day.

The bottles arrive in three different sizes for filling with one of eight or more different brands of beer, and the establishment centres round ten fully automatic assembly lines, each made up of a succession of interconnected machines. At the start, metal crates containing empties are unloaded from lorries and placed on one of several gently sloping runways, down which they slide into the Stores building, there to join a long but steadily moving queue on a power-operated conveyor. This conveyor is, in effect, a small scale railway, and at the head of the system is a marshalling yard comprising a veritable maze of sidings, branch lines, and junctions. Here, as a first and necessary step, the crates automatically sort themselves according to size—nips, small, or large.

The ultimate aim is to direct crates of a given size to one or more of the assembly lines, which run the full length of the main hall beyond, and this preliminary sorting is achieved by means of an ingenious arrangement of checks, stops, and starts, so contrived that a steady movement of traffic is at all times maintained without causing any piling up at vital junctions. Although this initial sorting of the crates is entirely automatic, the question

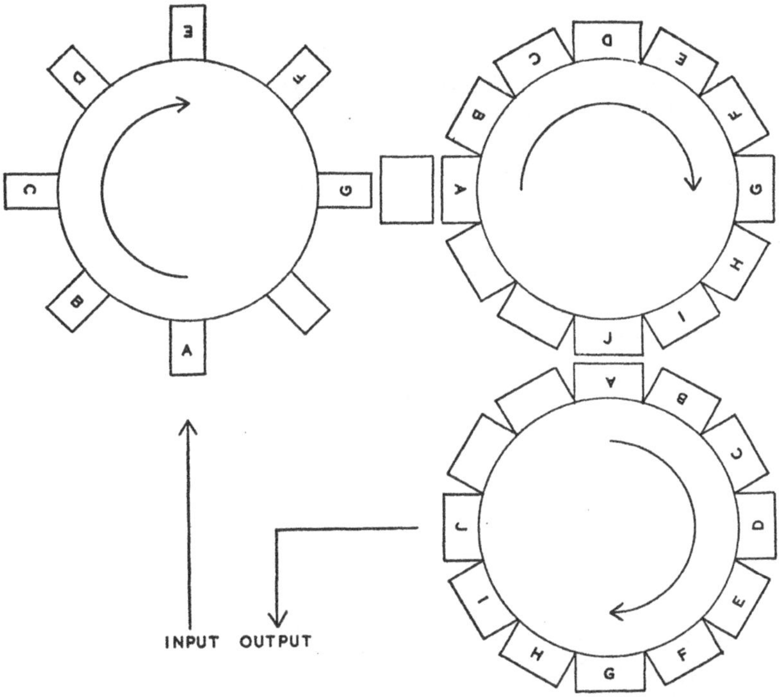

FIG. 11.—Operational cycle of the Hesser packaging machine. Essentially, the machine consists of three adjacent turntables, the synchronised movements of which are a series of rapid starts and stops. The first of these rotators has eight arms, which in size and shape exactly match the inside dimensions of the box-to-be, and it is about each one of these identical projections that a given container is progressively formed. First, however, cardboard is fed from a roll into another part of the machine, where, in preparation for what is to follow, it is slotted, glued, and cut off to length. The severed piece of cardboard is then shaped about one of the projecting arms at point *A*, after which, in quick succession, it is carried round to *B*, where the base of the now formed carton is closed; to *C*, where a glued paper wrapper is added; to *D*, at which point the bottom of the wrapper is folded; to E, at which station pressure is applied to the base in order to seal it; to *F*, and so on to *G*, where the completed carton, its top open, is ejected.

The entire process is continuous. That is to say, a carton which has undergone operations *A—G* is being followed by another which has experienced operations *A—F*, and this, in turn, has next to it a third carton which has passed through stages *A—E*—and so on, right back to *A*, where the initial folding of yet another carton is simultaneously taking place.

In the meantime, the carton which was ejected at *G* finds itself in a horizontal position in a tubular enclosure. Here, it is given a quick quarter turn, and is then pushed out of the tube and up-ended into a metal cup or holder, one of twelve attached to the rim of the second turntable. From point *A* once more, the carton moves round to *B*, and then to *C*, where a trip mechanism operates a weighing machine above, itself fed by hopper, which in turn is supplied by an endless conveyor belt. An accurately weighed portion of sugar or the like, thus released, starts on its way down a chute, and contrives to fall into the open mouth of the carton during its momentary halt at *D*. The loaded carton then goes on to *E*, to *F*, and to *G*, where a flat spring presses on the side of the carton, and brings the machine to a stop if a lack of resistance indicates an empty container. This question having been satisfactorily decided, the carton journeys on to *H* and to *I*, and so arrives at *J*, at which station it is transferred to the third and last of the turntables.

Again starting at *A*, the carton moves on to *B*, where the cardboard top is folded; to *C*, at which point the top of the paper wrapper is splayed out ready for glueing, and where, in addition, another check is made to ensure that the packet really does contain something (if, by some mischance, it does not, the empty carton will emerge from the machine with its top unsealed); to *D*, where, all being well, glue is applied to the wrapper top; to *E* and to *F*, where the final sealing occurs; and to *G*, *H*, *I*, and *J*, where heat is applied to dry the glue. The completed carton then leaves the turntable, and is carried by conveyor to a check weighing point.

At the refineries of Tate & Lyle, Limited, where a number of Hesser machines are in use, three operators are normally in attendance. Their task is to keep the machine in full production—to restart it, should it be brought to an automatic halt, to see that the supply of cardboard is replenished when necessary (every ½ to 3 hours, depending upon the size of the roll), to ensure that there are ample wrappers in the stack, and to maintain the level of adhesive in the trio of glue containers. Thus serviced, the mechanism will make, fill, seal, and deliver 2 lb. cartons of sugar at the rate of 75 per minute, or more than 65,000 in the course of a 14½ hour working day—and use up nine miles of cardboard in the process.

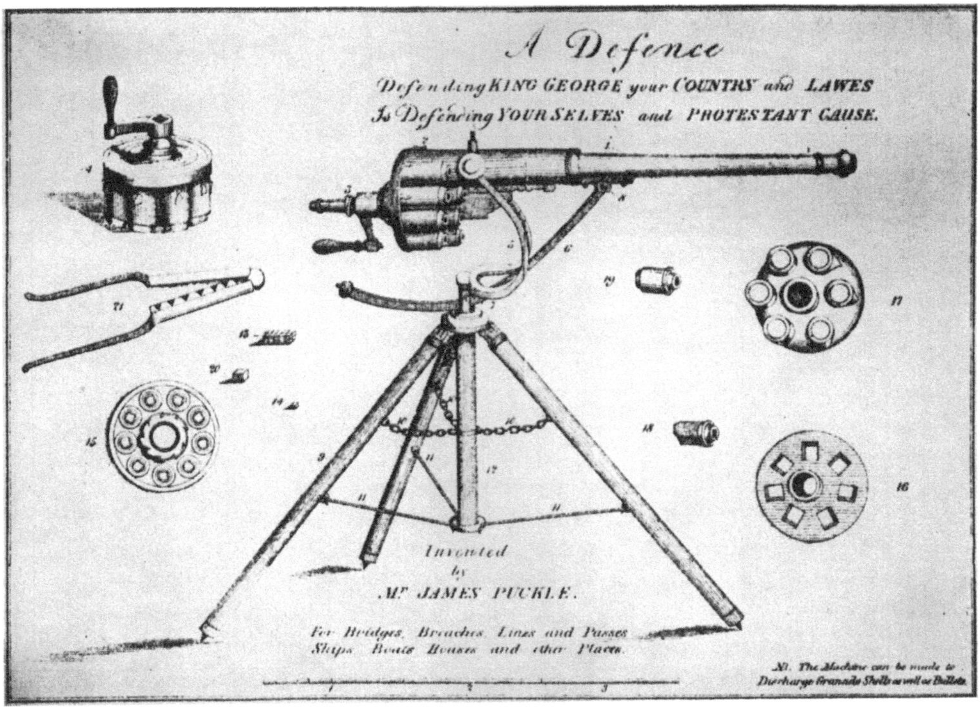

PLATE 12(*a*). The machine gun patented by James Puckle in 1718. In keeping with the sanctity of the times, it was designed to fire round bullets at fellow Christians, and square bullets at the Turks. Photo: *The Patent Office*. (*b*). A modern automatic carbine—the Neuhausen AK 52. The magazine holds 30 rounds, and the rate of fire is 400 rounds per minute. Photo: *Schweizerische Industrie-Gesellschaft*.

PLATE 13. The Skysweeper, a 75-mm. artillery machine gun designed to combat low-flying, high-speed aircraft. Radar and computer units are mounted integrally with the gun, which automatically spots, tracks and intercepts invading planes. Photo: *U.S. Army.*

PLATE 14(*a*). This machine makes a carton out of a sheet of cardboard, adds a printed wrapper, fills the carton with a weighed quantity of sugar, and then delivers the sealed package ready for use. Photo: *Fr. Hesser*. (*b*). A high speed automatic machine for the moulding and wrapping of butter and similar products. Its output is 80 packages a minute. Photo: *Kustner Frères et Cie.*

PLATE 15 (Picture 1). An endless succession of empty bottles is automatically sorted according to size. The bottles then find their way to one of ten production lines, where they are successively washed (picture 2), filled (picture 3), capped, Pasteurised, labelled (picture 4), recrated, and placed in store ready for despatch. (The frontispiece photograph shows the bottles being automatically marshalled to pass into the Pasteuriser.) Photos: *Ind Coope and Allsop Ltd.*

of their ultimate destination is decided by an operator who, sitting at a desk in the Empties Control Office, manipulates a panel of master switches in response to light signals from the various machines in the main hall. A green light indicates that an approach line is full to capacity, whereas a red one is an urgent call for more crates. The various streams of traffic, once they have sorted themselves out, are directed accordingly.

On arrival at the Bottling Hall, the crates are unloaded at one of the washing machines which stand at the head of each line. And while the empty crates continue their journey on the conveyor, the bottles are hand fed into the machine in batches, there to receive four separate washings before emerging for inspection —clean, sparkling, and bereft of all old labels. Any cracked bottles are removed manually, while the remainder, untouched by hand, move off in single file to a filling machine. Here, one after the other in quick succession, the empties are taken for a brief ride on a miniature merry-go-round, in the course of which gyration they each receive a measured quantity of chilled beer, piped to the machine from bulk storage. Thereafter, the filled bottles pass on to an adjacent capping machine, where crown corks, hopper fed into the device, are automatically applied and crimped.

The bottles now approach the Pasteuriser, where they make a right-angled turn and form up, still in single file, along its front edge. As soon as four dozen or so of them are thus deployed, a long bar, running the full width of the machine, gently pushes them forward *en masse*, until they find themselves on a slowly moving platform which, almost imperceptibly, carries them into the steaming interior. The push-bar, meanwhile, assumes its original position until another four dozen bottles are lined up, whereupon these, too, are herded into the Pasteuriser—and so it goes on.

The Pasteuriser is a large, box-like structure, with transparent observation panels in its sides, and measures some sixty feet in length. The bottles, 10,000 or so of them at a time, require no less than an hour to traverse this distance, and in the course of the journey the temperature is gradually raised to 60 degrees Centigrade (140 degrees F.), and thereafter slowly lowered.*

* Not all beers are so treated—in some, the fermentative process is completed after bottling, instead of prior to it. But the notion, voiced by

Out in the open once more, the bottles make their way to a labelling machine, where labels appropriate to their contents are deftly added and as quickly smoothed into place. They then pass on to an automatic packing mechanism, which arranges them into four rows of six (half pint size) or three rows of four (pint size) and holds them, thus grouped, until the arrival of an empty crate on the conveyor which runs immediately below, whereupon the cluster of bottles is dropped neatly and unerringly into the allotted compartments. The filled crate then heads for the Full Stores, and it may be that a container moving along an overhead track is destined for a line running below it. In this event, of its own accord, and precisely at the right moment, a hinged drawbridge, forming part of the upper track, descends so that one end of it rests on the line beneath, thereby enabling the approaching crate to slide down from one level to the other. The drawbridge then returns to its former position.

The operational activities of the various machines which go to make up the assembly line are all the while being noted by strategically placed recording instruments, which transmit a constant flow of information to a Central Control Room—the brain of the entire system. In this room, normally unattended, an imposing array of continuously clicking electrical counters show at a glance how matters stand—the number of bottles that have been filled, with what brand they have been filled, and the amount of stock of each that is held in the Full Stores, essentially a series of conveyor sidings, running the length of five different floor levels. At these Stores, not only is there a never-ending stream of fresh arrivals from the assembly lines, but there is also a constant despatch of crates to a fleet of waiting lorries. And as with the incoming empties at the beginning of the line, the control of outgoing crates is exercised from a room containing a panel of master switches. Here, an order is received, say, for eight two-dozen crates of Double Diamond, five of Allsopps

certain connoisseurs, alleged and supposed, that Pasteurised beers are of inferior quality because they are not fully fermented is, of course, fallacious. One of the advantages of Pasteurisation is that it results in a sediment free drink, easy to pour without the production of spectacular and unwanted cloud effects.

Stout, and three of Nut Brown Ale—and at the touch of a switch the required crates automatically leave their respective sidings, and converge upon one of the loading bays, for shipment to you and me.

It will have been noted that human intervention ordinarily occurs at two points only along the line—at the start, where the empties are transferred by hand from crate to washing machine, and at the finish, where the release mechanism controlling the fall of grouped bottles during re-crating is worked by foot pedal. Both of these operations could be undertaken by self-acting devices incorporated in the machines concerned,* and other contrivances (such as photo-cells) could take over the functions of the several observers, stationed at intermediate points, who are called upon to remove an occasional cracked or incompletely filled bottle from the line. Similarly, the continuous supply of sundry commodities used by the various machines, such as crown corks, printed labels, adhesive, and beer, involving the automatic replenishment of the already existing self-feeding arrangements, would present little difficulty. There remains the question of the overall direction exercised by the Empties Control Office, and by its counterpart in the Full Stores. This could be undertaken electronically, or the need for such supervision eliminated by the simple expedient of confining a given line, and its attendant conveyor sidings, to a single brand of beer, whereby a completely automatic bottling process could be achieved.

In the much more complicated automobile industry, an ever-increasing tendency towards automatisation is also to be observed. The final assembly lines in the new Austin building at Longbridge, each moving at a speed of seven feet a minute, are automatically fed from adjacent sub-assembly lines by way of an underground tunnel and some sixteen miles of conveyor. At the fourfold head of the assembly lines, there is a marshalling stores containing three-quarters of a million chassis parts

* An RCA machine rejoicing in the name of Full-Depth Uncaser and Washer-Loader was installed several years ago at the Hamilton Park plant of The Borden Company, in Chicago. It automatically removes empty milk bottles from their cases, and then feeds them to a washing machine.

(sufficient for two days), comprising nuts, bolts, oil pipes, and the like. Sets of these parts, stored in pallet containers, are marshalled on the moving tracks at regular intervals, chassis type vehicles at once being joined by a chassis frame, delivered by telpher. Thereafter, the arrival and positioning of the remaining major units is entirely automatic, the chassis (in the case of this type of vehicle) or the rear axle (where unit construction vehicles are concerned) making electric contact, by means of a plunger-operated limit switch, as it is carried along the track. Due notice is thus given of impending requirements, and as a result, engine and gear-box, and finally the body, fully trimmed and painted, are successively added at two prearranged points *en route*, self-operating hoists raising the various items up and over the line from conveyor sidings located in the tunnel below, and then gently lowering them onto the moving track.

The arrangement is such, moreover, that each assembly line can produce several kinds of vehicle—there may emerge a saloon or a van, with this or that design of engine under its bonnet, and with a left-hand or a right-hand drive. Thus, when a given set of components signals ahead in the course of its journey that it needs to be met in so many minutes by a power unit at the engine elevator, the particular type of engine required is determined by a sequenced programme. To this end, electromagnetically operated catches are incorporated in the system of endless conveyor chains, so that component-carrying carriages can be moved along or not, as required. By means of these indexers, as they are termed, any given engine can be selected from a so-called sequencing area, and from there directed to sidings adjacent to the tunnel hoists. Overall control is exercised electronically from a central point, all relevant data about each vehicle being codified on a punched card, and the card fed into a Hollerith machine. Guided by the series of electrical impulses thus produced, the integrated mechanisms do the rest: no matter in what order the various types of engines and bodies leave the sub-assembly lines, the correct component is delivered to the appropriate chassis at precisely the right moment.

But replete though they are with automatic mechanisms of all kinds, these final assembly lines nevertheless each require the undivided attention of 100 workers, with a further 7 in reserve (since, though the moving track pauses not, there are occasions when those who lavish attendance upon it must). Among other tasks, operatives are still called upon to marshall the chassis components; to assemble the front suspension and the prop shaft, the brake mechanism, and the clutch control; to fit the exhaust pipe and the road wheels; to secure the body, to fit the bumpers, and to position the seats; to supply water, air, and petrol; and to give a final inspection before the completed vehicle is driven off to the despatch bay.

It is true that in terms of physical effort, the operatives who perform these tasks are required to do little actual work, for power-operated tools are everywhere to hand—tools, incidentally, which need no lifting up or setting down, for they hang suspended in mid-air, readily to hand, to which position they return of their own accord on release. And while the automatisation of the final assembly lines is being completed, increasing attention will inevitably be given to the question of extending the process to sub-assemblies, and to the production of the components therefor. Considerable progress has been made in this direction already. Engineers of the Ford Motor Company, for instance, have devised a machine which not only assembles radiator caps, but which inspects the finished product, while among many ingenious mechanisms planned and built by Allied Steel and Conveyors, Inc., is a combined tyre mounting and inflating conveyor which first mounts automobile tyres on drop-centre steel rims (by means of a mounting head which accommodates itself to mixed sizes of tyres) and then proceeds to inflate them to any predetermined pressure—at a rate in excess of 600 an hour.

But although the manufacture of individual parts at present involves an extensive use of self-operating mechanisms, this is not to say that all that remains to be done is to link up existing machines with an assembly belt. Consider, for example, an automatic lathe producing a single item, such as brass nuts. Some supervision, however slight, will ordinarily be required, and its abolition is bound to introduce many complications. To ensure a continuity of supply without human intervention will demand not one, but several such lathes, not to mention a host of attendant mechanisms, necessary to feed and service them. Thus, from a store room, itself periodically replenished, containing ample stocks of extruded brass rod of hexagonal section, supplies will need to be fed continuously to one of two lathes located at the head of the production line. The emerging stream of nuts will at once pass on to an inspection unit (such as the Sigma*) for a multi-dimensional check, whereafter each component will take one of three routes, according to whether it is undersize (and hence scrap), oversize (and thus rectifiable), or correct. Any undersize material will join the brass waste from the lathes, and be taken by conveyor belt to a hopper for subsequent disposal. Oversize items will be diverted to a stand-by lathe, *i.e.*, to a machine other than the one which produced them, for corrective treatment, whereafter they will again present themselves for inspection. An inordinate number of rejects will indicate a need for resetting, and while this is being undertaken, production will be maintained by the second of the two lathes aforementioned. Automatic re-setting could doubtless be contrived progressively, interspaced by trial runs, until such time as the inspection unit signalled that nuts of the correct size were again being produced. And in this connection, it may be expected that the life of tools will be considerably lengthened in the future (as it has been greatly extended in the past) by the development of new and improved alloys. But even so, tools may break,

* In an actual installation, designed to deal with mass produced parts in considerable numbers (half a million a week), six hopper-fed Sigma machines are mounted on a single conveyor, divided into three sections. After passing through the gauges, rectifiable and rejected parts are deflected into separate containers, while accepted components pass into one of two bags at the end of the conveyor. When one bag is full, the stream of parts diverts itself to the other bag, thus enabling its companion to be removed and replaced. Should both bags become full, the flow of parts stops on its own accord, and a warning bell continues to sound until a full bag has been removed. Information about the number of parts submitted, accepted, and rejected, is provided by counting devices, which also indicate the number of times any of the dimensions is large or small. A complete analysis of the operation is thus provided.

CRAWFORD *Daily Sketch*

or, this failing, ultimately show such signs of wear that they will need to be replaced—contingencies which will certainly call for radical changes in machine design, if the necessary replacement is to be undertaken by the machine itself.

All this, be it noted, has been occasioned by the need to ensure the continuous production of a single item, the output of which is almost certain to be far in excess of the requirements of the primary product, with the result that the factory will either be in a perpetual state of unbalance as a whole, or subject to prolonged periods of idleness in part. At the present time, the economic aspect alone would suffice to condemn an arrangement so costly to install, and so wasteful to operate. But in the days ahead, such considerations may be expected to become a matter of little or no consequence, and there will in any case be nothing to prevent a sharing of any over-production by distributing it to adjacent assembly points, of which more anon. It will be obvious, meanwhile, that the greater the number of different component parts, the more complex will the problem of co-ordinating their production and assembly be. To offset this, the introduction of new and revolutionary methods and materials may be expected, as exemplified in the realm of radio by the replacement of sectional cabinets by one-piece, moulded cases, made of self-coloured materials; by the use of two dimensional printed circuits instead of three dimensional wiring; by the substitution of the minute transistor for the cumbersome thermionic valve; and so on.*

Such simplicative measures will be essential in cases where the number of individual components is normally very large. The modern automobile, for instance, is made up of some 7,000 separate parts, the manufacture and assembling of which by mechanical means alone would, at first sight, appear to offer almost insuperable difficulties. Yet, however remote the prospect may at present seem, the robot car factory is nevertheless attracting its due share of serious attention. In the course of a recent exchange of views between sixteen interested experts, one speaker expressed doubt that the automobile engine would ever be assembled by automatic means in his (unspecified) lifetime.

* In Chicago, at least two leading manufacturers of radio receiving sets have contrived to assemble their products from ready-made components with the aid of robot mechanisms—Motorala Incorporated, which concern is said to have installed a machine (known unofficially as 'The Monster') which regurgitates complete radios after consuming and digesting the constituent bits and pieces, and The Raytheon Manufacturing Company, whose automatic assembly line, operating under the aegis of two caretakers, has been reported to produce 1,000 sets daily.

At the time of this writing, much secrecy surrounds these establishments, and requests for detailed information were answered by polite declinations. The Raytheon Manufacturing Company, however, confirmed that it had done considerable work in the mechanisation of radio assembly lines, while Motorala Incorporated, in making a similar admission, also conceded an extensive use of printed circuitry.

To this, the answer was made* that it was not reasonable to expect machines to put together a highly complicated mechanism which had been designed in the first place to be assembled by men. No doubt, with patience, it could be done—but how very much easier it would be to re-design the engine completely, with a view to its being assembled solely by machines!

Thus it may be anticipated that the arrival of the automatic factory will be heralded by an extensive overhaul of existing manufactural procedure, assisted by a rigid standardisation of the component parts that are to be made and assembled. And the finished product, at any rate in the beginning, will no doubt be a strictly utilitarian affair, stripped of all non-essentials, and exhibiting many changes in recognised design. Possibly the first automobile to be made solely by automatic means, like a famous predecessor of earlier days, will be available in any colour the prospective owner desires, providing that the colour chosen is black. In as much as such a chromatic limitation had no adverse effect upon public demand for the Model T Ford (15,000,000 were produced in the course of two decades, during which period, thanks to new production methods and ever-increasing mechanisation, the price was halved, and then halved yet again) it seems reasonable to suppose that it will prove to be no deterrent in days to come. In any event, for a time at least, semi-hand made models will still be available for those who prefer them, even at forty or fifty times the price.

Sooner or later, the question of the making of major changes in the design of the products of automatic factories will have to be faced. Perhaps the making of essential improvements will be possible at first only by a ruthless scrapping of all the machines directly concerned. But in time, some measure of flexibility may conceivably be introduced, though to what extent this would require additional and more complicated machinery (and

* By E. W. Leaver, replying to R. E. Cross. For a full account of this discussion, reference should be made to 'The Automatic Factory' in the October 1953 issue of *Fortune*. The exchange of views took place under the chairmanship of Editor Eric Hodgins, and among those taking part were C. H. Patterson and G. P. Hitchings, of the Ford Motor Company, and R. J. Emmert, representing The General Motors Corporation.

whether, having regard for all the circumstances, it would be considered worthwhile) remains to be seen. In any case, it may be assumed that there will be no question of attempting to make allowance for a never-ending flow of variation orders, such as official supply departments are at present wont to inflict upon manufacturers of governmental equipment, particularly that of a military nature. Automatisation will leave little enough room for afterthoughts.

There remains the matter of the provision and processing of raw materials, and presumably existing methods and procedures will lend themselves to a gradual elimination of the human element, so that even an entirely self-operating steelworks can be envisaged as a possibility of the remote future. An important point here, as with component manufacture, will be the question of maintaining a balanced production, and a solution of the whole intricate problem of internal supply and demand seems likely to be found by arranging a judicious grouping of industries in terms of basic materials needed and component parts required (Fig. 12). Under such an arrangement, raw materials would arrive at a central reception point, A, for distribution to various processing plants, located in zone B, for conversion into usable commodities such as sheet steel, copper wire, extruded brass rod, plate glass, and so on. These materials would then be divided among the more numerous component-making establishments in zone C, the combined output of which would be shared by the even greater array of assembly plants in the outer zone D. Overall control, exercised electronically from the central point A, would be so contrived that the demands of zone D mechanisms would determine the rate of output of the zone C units, whose requirements would in turn dictate what quantities of materials each of the zone B plants produced.

In addition to external arrangements for the regular collecting of finished goods from peripheral storage points, and the periodic disposal of any unusable by-products and other waste, the scheme presupposes a continuous and adequate supply of power (*vide infra*) and raw materials. Some idea of what is already involved in this connection may be gained from the following tabulation, compiled from official sources of informa-

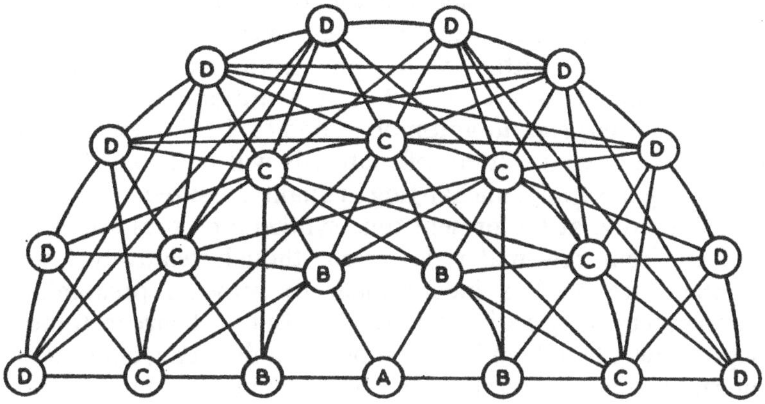

FIG. 12.—Automatic factories, grouped in terms of basic materials needed and component parts required.

tion, which lists a few of the many varied items which go to make up the weekly intake of the Austin factories at Longbridge:

Coal (tons)	1,400
Oil fuel (gallons)	33,000
Gas (cubic feet)	108,000,000
Water (gallons)	7,000,000
Timber (cubic feet)	7,500
Glass (square feet)	40,000
Paint (gallons)	12,000
Hides	2,500
Sparking plugs	17,500
Lamp bulbs	50,000
Ball and roller bearings	64,300

It will be clear that where multiple groups of fully automatic factories are concerned, in operation twenty-four hours a day, the provision of adequate supplies of such widely varying items would present no mean problem. Inevitably, the enterprise would be jeopardised by the uncertainties imposed by the need to mine ore, to grow timber, to raise cattle, and to cultivate cotton, and it may be expected that every effort will be made to render factories independent of all such dubious sources of supply.

Synthetic plastic materials are, in all probability, destined to provide an almost complete answer to this particular problem, ousting in the process commodities such as natural rubber, wool, porcelain, cotton, wood, and leather, and increasingly replacing metals wherever such a substitution can be successfully contrived, *e.g.*, in automobile bodies. In this way, the use of an unduly wide variety of materials will be avoided, and a continuity of supply, unchanging in quality, assured.

Admittedly, at the present time synthetic plastics are largely derived from raw materials which, though relatively few in number, are no less inconvenient to provide, predominant among them being coal and petroleum. But the benzene, toluene, napthalene, phenol, and other substances derived from coal and oil can also be synthesised from their elements, notably carbon, hydrogen, oxygen, and nitrogen. The parents of a whole host of plastic materials (including neoprene, perbunan, the polyvinyl resins, the poly-ethylenes, and methyl rubber) are the familiar hydrocarbons acetylene, ethylene, and benzene, of which the last two can, if necessary, be made from the first-named. Acetylene is thus a key substance in the preparation of many synthetic plastics, and one which can be formed direct from its constituents with the aid of intense heat (electric arc). A more convenient method of preparing the gas, however, is by the action of water on calcium carbide, the products of the reaction being acetylene and calcium hydroxide (slaked lime). Nor need the calcium hydroxide be wasted. When heated, it gives off water, and is thus converted into calcium oxide (quicklime), and this, if mixed with carbon (say in the form of coke) and heated yet again, produces carbon mon-oxide—and calcium carbide! Here, then, is a cyclic process by means of which a continuous supply of acetylene can be derived from coke and water (Fig. 13).

An example of the complexity of some of the many artificial substances which can be built up from such simple beginnings is provided by the super-polyamides. Acetylene, C_2H_2, if passed through a red hot tube, can be converted into benzene, C_6H_6, three molecules of the one polymerising (uniting) to form one molecule of the other. With hydrogen peroxide, benzene yields phenol (*alias* carbolic acid), and by the use of nickel as a catalyst, phenol can be hydrogenated to cyclohexanol, which in turn can be oxidised to tetramethylene dicarboxylic acid. And

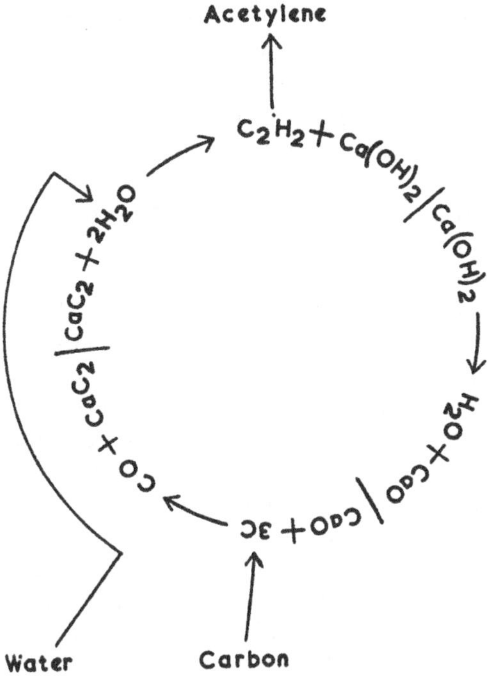

FIG. 13.—The acetylene cycle.

(if you will bear with me for a moment longer) the interaction of this substance with hexa-methylene diamine, itself a derivitive of the aforementioned tetra-methylene dicarboxylic acid, produces polyhexamethylene adipamide, better known as nylon.

Clearly, it will be far easier to provide a constant supply of a few raw materials such as water and coke, than to bring together a multitudinous array of widely differing commodities ranging from metalliferous ores and potter's clay to cow hides, rhinoceros horn, and hardwood logs. The tendency will thus be for all such substances to make way more and more for synthetic substitutes, it being borne in mind that the man-made material need not necessarily differ in composition from its natural counterpart (*cf.* artificial indigo). But although synthetic materials seem destined completely to replace animal and vegetable products as at present used in clothing, furniture, buildings, and the like, there is plainly a limit to the extent to which they can be expected to act as a substitute for metals. Where high tensile strength is needed, or electrical conductivity is required, some use of metals will be essential—and pure metals are elementary substances, incapable of ordinary (though not, of course, of nuclear) synthesis.

It does not necessarily follow, however, that a laborious mining of ores will thus be unavoidable. It so happens that there exists a source of mineral wealth, as yet virtually untouched, which is freely available to all: the sea. Untold millions of tons of metals are contained therein—iron, gold, copper, silver, zinc, magnesium, uranium, *et al*, one and all to be had for the taking. Here, indeed, is a superabundance of metals, a plentitude so diluted, however, that hitherto it has defied economic extraction. This hindrance, *i.e.*, the lack of a sufficiently cheap source of power, will surely be overcome before long, and in the days ahead a pipeline to the ocean may well bring all the raw materials of which any manufacturing centre has need.

In the meantime, the gratifying result of providing a succession of automatic mechanisms with an uninterrupted flow of raw materials is already to be seen at the Harworth premises of Glass Bulbs, Limited, to which enterprise, reference has already been made. Here, two production units, each about the size of a railway locomotive, are housed side by side in solitary splendour. The entire factory is virtually built round these two machines, which are tended by an imposing array of subsidiary robot mechanisms, dedicated to the task of keeping pace with a voracious and insatiable demand for molten glass. To meet this demand, supplies of dolomite, feldspar, limestone, sand, and soda ash periodically arrive by rail at the base of a hundred foot high octagonal tower. From top to bottom, seven of the eight bays of this huge structure are in the form of silos, each capable of holding 1,500 tons of material, and within which the various ingredients are stored after being conveyed aloft by suction pipe and other means.

From their respective silos, the raw materials pass down chutes to automatic weighing machines, and then enter a rotary mixing drum. Remote, push-button control of this operation is exercised by a staff of two, housed in the eighth bay of the tower, whereafter the blended mix is loaded into travelling canisters and carried on a roller runway to a self-operating hoist, and so reaches a hopper from which it finds its way into a central furnace. This is of the regenerative type, gas fired, and kept in operation continuously day and night. Its maximum output is 160 tons of refined glass every twenty-four hours, which vitrescence pours from a forehearth in a continuous, white-hot stream. Thus fed, the two machines are together capable of producing 1,400 glass bulbs every minute, or 84,000 every hour—*more than 2,000,000 every day!*

An evident desideratum of the automatic factory will be a steady demand for its products, and up to a point this will be ensured by the very provision of plentiful supplies, available to all. It seems likely, however, that certain items will remain in small demand, even as a gift, and it may be expected that grandfather clocks, boomerangs, rocking chairs, barrel-organs, fontal basins, battleships, and other such superfluities of the age, will continue to be custom made until such time as they share the fate of the patchwork quilt, the warming pan, the snuff box, and the arquebuse.

II

Aspects and Implications

MANY of the new and revolutionary labour-saving devices which were introduced during the 17th and 18th centuries—Hargreave's spinning jenny and Arkwright's water frame among them—were attacked and destroyed by the very workers whose arduous toil these mechanisms were intended to relieve. Nowadays, such senseless sabotage is rarely a matter of howling mobs and pounding sledge hammers. Rather does it find expression in a more subtle fashion—in union insistence, say, that a contrivance which can do the work of two men shall be run at half its normal speed, or that a Diesel locomotive which has no need of a fireman must, nevertheless, carry one, and so on and so forth. The Luddite mentality, in brief, still lingers on, despite the fact that convincing evidence of the short-sightedness of such an outlook is not far to seek. It is no accident, we may be sure, that where inhibitions about the use of machines are fewest, there the general standard of living is to be found at its highest. Aided by extensive mechanisation as well as by more accessible seams, the output of the American coal miner is more than five times that of the British underground worker. In the U.S., indeed, it appears to be tacitly accepted that because of the high price of labour, it is incumbent upon the manufacturer to bring down his costs in any way that he can, even at the expense of some slight inconvenience to the worker. But elsewhere, a less enlightened attitude not infrequently prevails. Before me as I write is a report* of an American machine, imported into this country at considerable expense by a well known aircraft firm in an attempt to speed up production, which has stood idle for no less than nine months (and which may well be standing idle still) because union officials cannot agree which particular

* Published in the London *Daily Express* for 5 June, 1953.

PLATE 16(*a*). This machine, devised by the Industrial Rayon Corporation of America, performs the entire process of converting liquid cellulose into finished rayon yarn. The picture shows a bank of processing reels, down which the yarn travels in an unbroken length from the spinning jet, receiving various chemical treatments as it does so. Photo: *Courtaulds, Ltd.* (*b*). A general view of the Bartrev Press, which converts resinated wood waste into compact building board, cut to length from a continuously formed strip. Photo: *The Vere Engineering Co. Ltd.*

PLATE 17. One man controls the Austin final assembly lines. Unit sets of minor components are placed on the moving tracks at regular intervals, those relating to chassis type vehicles being at once joined by a chassis frame, delivered by telpher. Thereafter, the arrival and positioning of other major units is entirely automatic, engine and gearbox, and finally the body, lowering themselves into position. The radiator is then filled, petrol is pumped into the tank, and the vehicle is driven away. Photos: *The Austin Motor Co. Ltd.*

worker shall be permitted to press the button necessary to set it in operation—one of the insuperable difficulties being, apparently, that the machine performs the work, not merely of five men, but of five men of various trades!

The question of redundancy is, of course, at the root of all such imbecile situations. On the one hand there seems to be a reluctance to face the fact that machines are in many respects superior to men, and on the other an almost pathological fear that this superiority may ultimately result in the displacement of the worker altogether—as is more than likely. But hostility towards mechanical innovations on this score and at this late date is a gesture about as futile as it is absurd. And not only is the process of displacement well under way, but it is actually being hastened by the irresponsible attitude of some of the workers themselves. Several managements have confessed that they would gladly dispense with the services of their labour force in its entirety, were the necessary mechanisms immediately available.

But that day is not yet. Indeed, even the often expressed conviction that the introduction of labour-saving machinery would lead to widespread unemployment has, so far, been consistently disproved by events. No doubt the advent of the horseless carriage was a serious blow to the stable hand and to the blacksmith, and to all those dependent for their livelihood on the continued use of *Equus cabalus*. But at the same time it has given rise to an enormous extension of employment in other directions —in automobile factories and numerous subsidiary concerns, in public and private transport undertakings, in garages and filling stations, and in oilfields and refineries, not to mention all the incidental work it has provided for the politician and the civil servant, the policeman and the lawyer, the surgeon and the undertaker.

As a case in point, consider the employment figures of the Austin Motor Company, obligingly supplied from archival sources. In 1910, the number of the Company's employees was 1,800. By 1950, despite a policy of progressive mechanisation, this figure had grown to no less than 19,500 (Fig. 14). Thus, far from leading to a loss of jobs, an ever-increasing use of machines

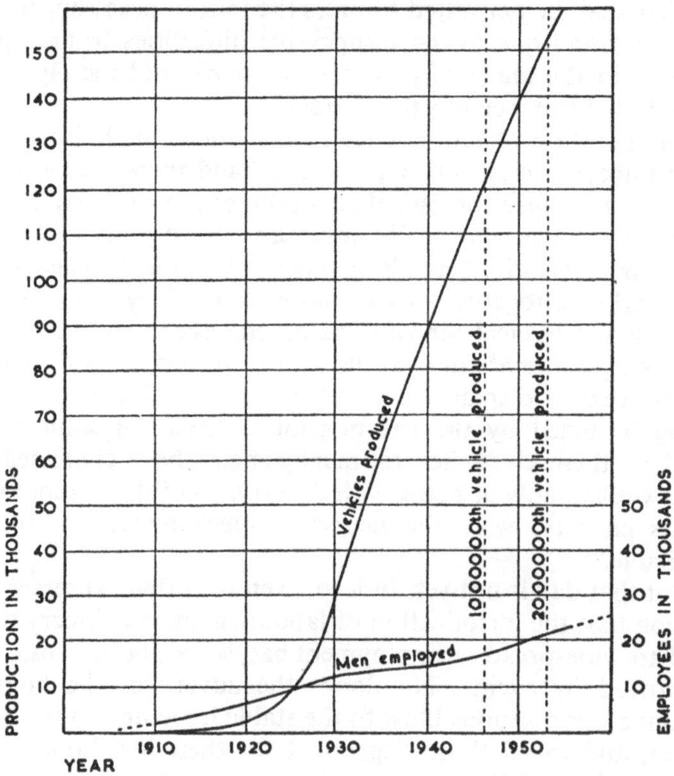

FIG. 14.—Austin production chart. Note that whereas it required forty years to complete the one millionth vehicle, only seven years were needed to produce the second million.

was in fact accompanied by a ten-fold rise in the number of workers required to operate and service them. But this, of course, is not the whole of the story. During the same period, the number of vehicles produced leaped from a modest 600 a year to an astronomical 150,000—an increase of two hundred and fifty-fold! In other words, the ratio of workers employed to vehicles produced, steadily decreased throughout the four decades under review, falling from 3 : 1 to 0·13 : 1. Relative to output, that is to say, manpower requirements have dropped to such an extent that it may well be enquired how long it will be before unlimited production is achieved without the aid of any employees at all (Fig. 15).

This question, and the circumstances which prompt it, are by no means peculiar to the automobile industry, and two distinct tendencies are inherent in the existing situation, in which an as yet unsatisfied demand for consumer goods leads to the employment of more and more workers, whose ever-growing output is all the while being still further increased by mechanical aids. Matters may continue thus for a considerable time to come—but it will be evident that they cannot be expected to go on indefinitely and without end. Sooner or later, at various indeterminate dates in the future, the saturation point in respect of one commodity after another will at last be reached, and thereafter the high rate of production which made possible this satiation will no longer be required. The extent of the subsequent fall in output will be governed by replacement needs, and these should provide a steady, if much diminished, demand. And in meeting this, it will be plain that any further mechanisation, if by this

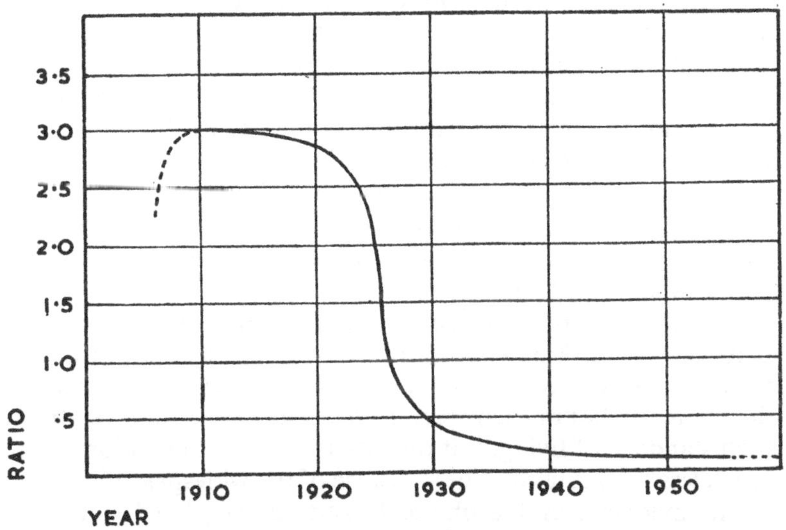

FIG. 15.—Austin output. The continuous decrease in the ratio of workers employed to vehicles produced from 1910 onwards will be evident.

time the factories concerned are not already fully self-operational, can occur only at the expense of the worker, until in the end he is virtually eliminated from the manufactural scene.

This, I suggest, is a prospect which can be viewed with equanimity by all concerned. The consumer will obviously have nothing to lose. And from the manufacturing point of view, there is everything to be said for a production unit which has no need of canteen facilities and other social amenities; which requires no time off for recreation and recuperation; and which does not fall sick, or arrive late, or indulge in restrictive practices, or go on strike, often because of some trivial excuse or imaginary wrong. As for the much maligned worker, he will surely make no complaint about being relieved of the compulsion to toil for a living, if such relief is unaccompanied by material want and economic distress. Few coal miners, it may be supposed, pursue their dangerous and unhealthy calling purely for the love of it, or even out of regard for such bodies as the National Coal Board. And fewer still, it may be imagined, would hesitate for one moment to throw up the job to-morrow, if they could safely do so without fear of the consequences to their families and themselves. As the figures for voluntary absenteeism in the pits go to show, it is plain that the high wages which British miners now receive do not provide a satisfactory answer to the problem of how to increase the output of coal. No underground worker who can satisfy all his immediate wants by labouring for four days out of seven, can really be expected to hazard life and limb by working a five-day week—and pay income tax on his additional earnings for the privilege!

The tendency here evident is a general one, as a comparison between the working conditions of a century ago with those of to-day at once reveals. Not only does the artisan now need to devote fewer hours each week in order to maintain a relatively high standard of living, but the intensity of that devotion is also much less than it used to be. Because of the widespread influence of the machine, undue physical exertion, *i.e.*, hard work, has largely gone out of fashion, even where there is still some call for it. In the old days, any bricklayer worthy of the name would lay up to 1,000 bricks a day, and think nothing of it. His modern counterpart (at any rate in Great Britain) is content to reach less than half this total, and no doubt considers that he has earned his keep. And gone with the habit of hard work is much of the old pride of craftsmanship—and understandably so. What man can be expected to take an abiding interest in the making of a component which pours out of a machine by the thousand while he stands idly by, his only concern the possibility that some improbable breakdown may occur? The role of the so-called worker is fast degenerating into that of a mere looker-on, a demoralising state of affairs which can best be remedied by a hastening of the day when the machine will be enabled to take over completely.

There is thus the prospect of unemployment on a vast and unprecedented scale among large sections of all highly industrialised peoples, albeit this enforced leisure will not be clouded by any lack of material needs. In the Western hemisphere, the emergence of the automatic factory may be expected to lead to an eventual collapse of the price system, with a benign form of communal ownership taking its place—an unlooked for outcome which will surely constitute the supreme irony of the times. As for this, it seems to me that communism—'a theory or condition of things according to which private property should be abolished, and all things held in common'—has long been a feature of certain revered institutions in our democratic midst. In the Armed Services, for example, where conscripts are fed, clothed, housed, and transported by a munificent authority which also provides all the equipment necessary for the pursuit of their calling. Here, indeed, is communism, if not at its best, then most certainly not at its worst. And in an age in which the State tends increasingly to assume responsibility for the welfare of the citizen, *i.e.*, to encroach more and more upon his liberties as an individual, the final outcome is not much in doubt. In the days ahead, no government will find itself able indefinitely to resist the temptation to acquire control of the means of automatic production. And with the output of goods steadily keeping pace with demand, the necessities of life, together with many of its luxuries, will be freely available to all. Mere money, that is to say, promises to become a meaningless token, if not something of an actual encumbrance. Its eventual discardment should serve to simplify life considerably—and at the same time add greatly to the ranks of the jobless by consigning thereto innumerable cashiers, almoners, book-keepers, treasurers, pay-masters, insurance clerks, shop assistants, bursars, stock brokers, financiers, bankers, money lenders, and pawnbrokers, along with hordes of minor officials, hitherto intent upon a sedulous collecting of rents, rates, fares, fees, fines, dues, imposts, tithes, taxes, and tolls.

As this progressive disemployment inexorably proceeds, increasing urgency will attach to the question of how the leisured masses are to occupy their time. At first, I dare say, the problem will be solved in part by an equitable sharing of such work as remains to be done, and by the introduction of a one-day week, or even a two-day month. And instead of only two or three men doing the work of one, as at present obtains, no doubt a dozen or more will be so engaged. On the other hand, as the available tasks become fewer and fewer, the possibilities of joint participation will steadily diminish—consider the noological obstacles in the way of transforming a one-time shop steward into a professor of Greek history, or of converting an ex-docker into a practising gynæcologist, or of turning an erstwhile reader of gas meters into a nuclear physicist!—until in the end vast numbers of the populace will be left with little or nothing to do at all. Those among the untutored multitudes who evince both an inclination and an ability to learn may expect to receive every encouragement to achieve a useful status, while the remaining two thirds (or four fifths, or nine tenths) will, perforce, be left more or less to their own unimaginative devices.

What is here betokened is an ordered diminution of the population, in a deliberate attempt to achieve quality—by eschewing mere quantity. Such a programme will, above all, need to avoid the appalling butchery which has been a feature of so many depopulative measures in the past, from the periodic slaughter of the helots by their Spartan masters to the merciless liquidation of the Kulaks by the Soviets in the 1930's, and the even more recent extermination of German Jews by the Nazis. Incredibly enough, despite all such mass executions, and

LANGDON *Lilliput*

notwithstanding the added toll of war, famine, and disease, mankind has not merely managed to survive, but has actually emerged from its long ordeal more numerous than ever before. In the past 150 years, the population of the world has more than doubled, rising from 900 to over 2,000 millions, and the rate of increase is now so high that the statisticians have at last begun to sound notes of alarm—one expert has just warned us that if the present trend continues, in another 2,000 years the congestion will be such that there will not be room for people to lie down. This predicament promises to be particularly inconvenient, as by that time most of the populace may well be chronic invalids, the degenerate descendants of unfit ancestors who were enabled to beget others of their kind thanks to the undue prolongation of their lives by a misguided application of medical science. Is it really desirable that sufferers from inheritable diseases should be kept alive, so that they may bequeath their affliction to one or more unsuspecting offspring?

In an irresponsible age such as ours, which moreover tolerates unrestricted propagation among the feeble-minded, drastic reform in these vital matters is plainly long overdue. The very least that is required is the passing of legislation designed to limit the size of families in strict accord with the physical condition and the I.Q. rating of their would-be parents—with compulsory sterilisation the penalty awaiting all those who choose recklessly to ignore the regulations. Having regard for the fact that between the years 1750 and 1900 the inhabitants of the now over-crowded British Isles increased in number from an estimated six and a half to more than forty millions, it may be safely assumed that a reversal of the process could be brought about within a reasonable period of time—if not entirely in accord with natural law, then at any rate humanely, and without recourse to the gas chamber and the extermination squad.

Since the political implications of such a eugenic enterprise could hardly be ignored, all this presupposes an agreed course of action, to be undertaken in concert by the technologically advanced nations of the world. And while the surviving exponents of mediocrity continue their subsidised existence in metropolis and kraal, performing their immemorial rites, worshipping their imaginary gods, and, perforce, maintaining a reproductive rate designed to ensure their eventual extinction, there will be introduced among the more intelligent minority of mankind an intensive programme of selective breeding, in a belated effort to improve the human stock. To this end (bearing in mind Galton's famous dictum that the power of nature is far stronger than the power of nurture), recourse will be had to artificial insemination, that an acknowledged genius may be enabled to father ten thousand experimental offspring without undue embarrassment or fatigue, a technique which, in the feminine interest, will no doubt be replaced in time by ectogenetic processes carried out in the aseptic surroundings of a governmental hatchery. So far as the male parent is concerned, of course, the number of progeny might just as readily be ten million as ten thousand. For every child that is sponsored in the time-honoured manner, upwards of five million potential brothers and sisters are left to perish ignominously in a uterine death-trap—such is the wasteful and improvident process which a careless and indifferent nature has devised for us. A scientific approach to human conception will not only put an end to such incredible prodigality, but, thanks to the deep freeze methods of indefinitely preserving spermatic fluid which have been developed, it will enable a man to achieve fatherhood a thousand years or more after he is dead.

That this vital research will inevitably result in failures as well as in successes, it goes without saying—with a selected sire, there is always the risk that a detrimental gene will escape early detection. But a careful check will be kept on each new breed that is produced, and any strain which in later life developed criminal tendencies, or a leaning towards racialism, or a predilection for theology, or exhibited some other equally undesirable trait, would be discontinued forthwith in the general interest. And so, by what will begin as a process involving considerable trial and not a little error, the foundation will be laid of a cosmopolitan intelligensia, owing allegiance to no particular race or creed, speaking a common tongue, and imbued with identical aims and aspirations: the pursuit of knowledge, and an intelligent application thereof.

III

Power Without Precedent

A PREREQUISITE of any robot era is an unfailing power source and a prime mover which is itself entirely automatic in action. As it happens, this dual requirement is neither as visionary as it seems, nor as extravagant as it sounds, for it has long since been met. The Iowa Electric Light and Power Company, of Cedar Rapids, in the U.S., for example, has had an automatic hydro-electric station continuously in operation, without anyone permanently in attendance, for more than thirty years!

In the past, however, despite its manifest advantages, water power has remained neglected and comparatively little used. For one thing, suitable heads of water are seldom available precisely where they are most needed, and the stupendous Victoria Falls, over the brink of which (according to season) from 3,500,000 to 75,000,000 gallons of water pour every minute, continue to run to waste because of their location in the midst of the African bush.* For another, the creation of suitable conditions reasonably near to populous areas, by a judicious programme of dam construction, calls for vast engineering projects and a heavy outlay of capital. And for a third, there has to be taken into account the inherent laziness of man, and his instinctive inclination to take the line of least resistance.

Thus it is that we live in a spendthrift age which chooses to rely mainly upon coal and oil as its source of power—to the detriment and to the discredit of us all. To our detriment, because our methods of fuel consumption are on a level, hygienically,

* When the author visited the Falls in 1949, he learned that a small (2,000 kW) hydro-electric station had been established there about ten years earlier for the benefit of nearby Livingstone, a town of some 1,500 (white) inhabitants. Like its counterpart at Cedar Rapids in the U.S., the station is entirely automatic in operation.

with the once popular custom of unceremoniously discharging the contents of household utensils (not necessarily of a culinary nature, as a well known portrayal by Hogarth serves to show) from an open window into the street below—compare our passive acquiescence in the existence of millions of belching chimneys, unendingly pouring their lethal fumes into the air we breathe;* and to our discredit, because this contamination arises from a wanton destruction of irreplaceable stocks of a valuable raw material which, by hydrogenation and other means, could have been made to yield untold wealth in the form of naphthalene, ethylene glycol, toluene, aniline, aspirin, and countless other useful products too numerous to mention.

And as if this were not enough, an acute shortage of coal now impends. Although known reserves, even after making allowance for increased consumption in the future, are sufficient to last for more than 1,000 years, the actual output of the mines is hard put to keep pace with an ever-growing demand. This ominous trend is particularly evident in Great Britain, which holds the unenviable record of using more coal per head of the population than does any other country, and which has already been reduced to the humiliation of having to bolster home production with foreign supplies. As little as twenty years hence, these home needs (now running at some 205,000,000 tons per annum) will have risen considerably above their present figure, and there appears to be little or no prospect of their being fully met.

What is the remedy for this sorry state of affairs? As I write, desperate attempts by the National Coal Board to provide more coal are being accompanied by official assurances in the House of Commons that every effort will be made to minimise the poisonous effects of burning it—*i.e.*, there is to be less smoke, though more of it. I submit the novelty that the only effective way to stop polluting the atmosphere is to stop polluting it, and

* According to Dr. A. Parker, Director of Fuel Research with the Department of Scientific and Industrial Research, some 2,500,000 tons of smoke and grit, along with 5,000,000 tons of poisonous sulphur di-oxide, are discharged into the air over the British Isles annually from the burning of coal alone.

suggest that as a first and decisive step, there should be a world-wide renunciation of the use of coal as a source of heat—the ban to come into force, say, in ten years from now. This decade of grace (which could be extended if necessary) would enable due preparation to be made: the carrying out of a widespread programme of industrial and domestic electrification; the conversion of existing power stations from coal to oil burning; the provision of additional generating facilities; the institution of an intensive search for new oil-fields; and so on.

The proposal has the merit* that while it recognises the impracticability of abandoning the burning of coal and oil simultaneously and forthwith, it substitutes electricity for the one at the expense of the other—it being acknowledged that the unprecedented demand for petroleum products thus created would unduly hasten the exhaustion of available supplies. In the meantime, there would be ample opportunity, first to supplement, and ultimately to replace entirely, the output of oil-burning power stations with electric current generated by other means, thus finally ending atmospheric pollution once and for all. As for the effects of an eventual oil famine upon locomotive devices using portable heat engines, it seems a reasonable assumption that when the time arrives, this will have been anticipated in a number of ways—*e.g.*, by the production of synthetic fuels and propellants such as benzene, alcohol, hydrazine, and hydrogen peroxide, and by inventions which will make available electrical energy by way of light-weight, quickly-charged storage batteries, or from central generating stations beaming power by radio, or by other means.

All this, it will be apparent, assumes the successful production of electrical energy on a scale not hitherto attempted, and by methods as yet untried or little used. Precisely how is this to be

* And the added virtue, in the case of oil-deficient countries such as Great Britain, that economic necessity would serve to intensify efforts to establish an independent source of power, thereby making some amends for past neglect. As to this, more than 10,000,000,000 tidal horse-power has continued to go to waste round our coasts while one Commission after another has solemnly debated the advisability of making some use of it—including an estimated saving of 2,000,000 tons of coal a year by means of the long-proposed River Severn scheme alone.

accomplished? A few years ago, the answer would have been confined to a listing of long advocated and potentially useful power sources such as the sun's rays, the tides, and the internal heat of the earth—recourse to one or more of which may well still be made. But recently, a new and revolutionary power source has unexpectedly come to the fore—nuclear fission, by virtue of which the energy release of one pound of uranium is equal to that produced by the burning of nearly 4,000,000 pounds of coal!

The story of this momentous achievement goes back to nearly a century ago, when it was first noticed that salts of the metal uranium could affect a photographic plate in the dark—a phenomenon subsequently examined by Becquerel in 1896, with the discovery of the rays named after him, and more picturesquely termed *lumière noire* (black light) by le Bon. It was later established that this mysterious radiation, under the influence of a magnetic field, could be split up into three distinct groups, designated alpha-, beta-, and gamma-rays. Of these, it transpired that the gamma-rays were akin to X-rays (and hence a true radiation), while the so-called alpha- and beta-rays turned out to be minute particles in the guise of positively charged helium atoms and negatively charged electrons respectively. So began a painstaking investigation into the nature of matter from which there gradually emerged a theory of atomic structure which has since found triumphant confirmation in the release of nuclear energy and the explosion of the atomic bomb.

The essence of the situation is that all matter—from holy oil to gutta-percha, and from hippopotami to demerara sugar—is made up of one or more elemental substances (92 in number) whose atoms can in turn be resolved into various arrangements of three component particles:

Particle	Mass	Charge	Symbol
Proton	1·0079	+1	${}_{1}H^{1}$
Neutron	1·0085	0	${}_{0}n^{1}$
Electron	0·0005	−1	${}_{-1}e^{0}$

It will be observed that the mass of the electron, compared with that of the proton and the neutron, is negligible, and that while protons and electrons carry a unit charge (of opposite sign), neutrons, as indeed their name is intended to convey, are

SABO *Cavalier*

I wouldn't sit on that if I were you, Mac!

electrically neutral. These facts are duly expressed by the corresponding symbols, the number below the line denoting charge, and that above it indicating mass.*

Atoms comprise a comparatively heavy nucleus (exclusively the domain of protons and neutrons) surrounded by one or more satellite electrons, and the arrangement is such that the association as a whole is normally electrically neutral. The hydrogen atom, for example, consists of one proton and one electron. The simple manner in which this equilibrium of charge is progressively maintained, and the close relationship it bears

* To the confusion of the uninitiated, nuclear physicists do not bother to distinguish between a neutral hydrogen atom, and a hydrogen atom stripped of its electron, *i.e.*, a positively charged proton. Similarly, an identical symbol is used to denote both a neutral helium atom and a doubly charged alpha-particle, *i.e.*, a helium atom minus its two electrons. Strictly, a proton should be designated $_{+1}H^1$, and an alpha-particle $_{++2}He^4$.

to an all-important elemental characteristic known as atomic number, will be apparent from the tabulation hereunder:

Element	Atomic number	Protons in nucleus	Attendant electrons
Hydrogen	1	1	1
Helium	2	2	2
Lithium	3	3	3
Beryllium	4	4	4
Boron	5	5	5
......	...	...	...
Uranium	92	92	92

But whereas an atom of lithium which sustained the loss of a nuclear proton would be transformed from an element of atomic number 3 into one of atomic number 2, the neutron content of an atom may vary (within certain limits) without in any way changing the identity of the element concerned, though not without affecting its mass. An extreme case of such mass modification is provided by the doubling which occurs when a neutron is added to the single proton of an ordinary hydrogen atom, with the production of an atom of deuterium, *alias* heavy hydrogen:

$${}_1H^1 + {}_0n^1 \rightarrow {}_1H^2.$$

So-called heavy water (which has a density at 25 degrees C. of 1·106, as against 1·000 for ordinary water) is of course water in which hydrogen has been replaced by deuterium. The possibility of a semi-heavy version will be noted:

$$H{-}O{-}H \qquad H{-}O{-}D \qquad D{-}O{-}D.$$

Elemental variants such as deuterium are termed isotopes, and examination has revealed that nearly all the natural elements are made up of a mixture of two or more of them. Thus, no less than three different kinds of silicon atoms are found in ordinary sand, each of which has 14 nuclear protons in association with either 14, 15, or 16 neutrons. There is, however, a limit to the number of neutrons which can be added to, or taken away from, an atomic nucleus without causing instability. Thus, two other silicon isotopes, containing 13 and 17 neutrons respectively, have been produced artificially in the laboratory, and each represents a configuration which is subject to spontaneous decay. In the heavier isotope, the surplus neutron breaks down into a proton and an electron, and the proton thus formed is retained in the nucleus, while the electron is emitted. As a result, the nuclear charge is increased by 1, and the atom of silicon thereby becomes an atom of phosphorus:

$${}_{14}Si^{31} \rightarrow {}_{15}P^{31} + {}_{-1}e^{0}.$$

In the case of the lighter isotope, on the other hand, it is a proton which suffers breakdown, giving rise to a neutron and a positron, a particle analogous to the electron, but carrying a positive charge. This positron is ejected, whereby the atom of silicon is transformed into an atom of aluminium:

$${}_{14}Si^{27} \rightarrow {}_{13}Al^{27} + {}_{+1}p^{0}.$$

Most of the elemental isotopes found in nature to-day exhibit atomic stability, a circumstance which need occasion no surprise, as any unstable forms have had billions of years in which to achieve equilibrium since the occurrence of the cosmic upheaval which gave them birth. But among the heavier elements, from thallium (atomic number 81) onwards, radioactivity is the rule rather than the exception.* In the case of uranium, the slow, natural decay of this substance, which has a half-life period of some 4,500,000,000 years (Fig. 16), is characterised by the emission of a succession of alpha-particles and so-called beta-rays (electrons), the end product being a stable isotope of lead. At the other end of the scale, numerous artificial radio isotopes are known whose half-life period is reckoned in days, hours, minutes, and even seconds. That of natural radium is 1,590 years, a time sufficiently short for all supplies now existing on earth to become virtually extinct in about 25,000 years. The reason why the element has not vanished long ago is explained by the fact that it is one of the intermediate products of uranium

* It is remarkable that feeble radioactivity is exhibited by the potassium isotope of mass 40 (K 40) and by the rubidium isotope of mass 87 (Rb 87), both of which occur naturally.

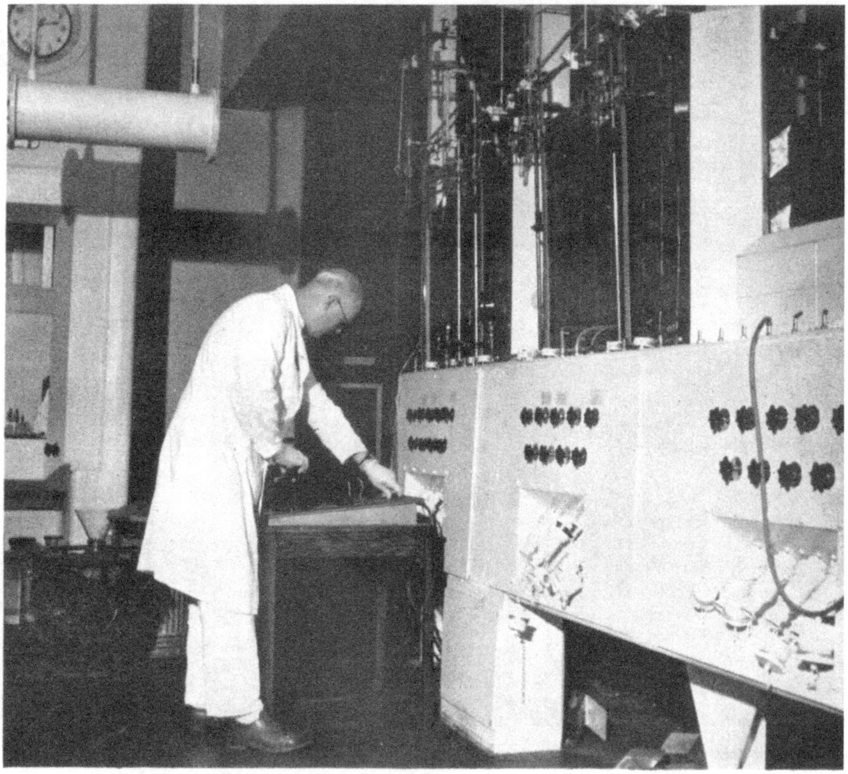

PLATE 18(*a*). In this radiochemical laboratory at the Atomic Energy Research Establishment, Harwell, plutonium is extracted from irradiated slugs of uranium. The process is carried out behind a thick lead wall and is controlled from outside the shielding with the help of a variety of instruments. Photo: *The Central Office of Information, London*. (*b*). The manipulation of radioactive materials with the aid of mechanical hands and three-dimensional television. By the use of polarising filters and polarised viewing spectacles, the dual images shown on the television screen are made to appear as one three-dimensional image to the operator. Photo: *Argonne National Laboratory, Chicago*.

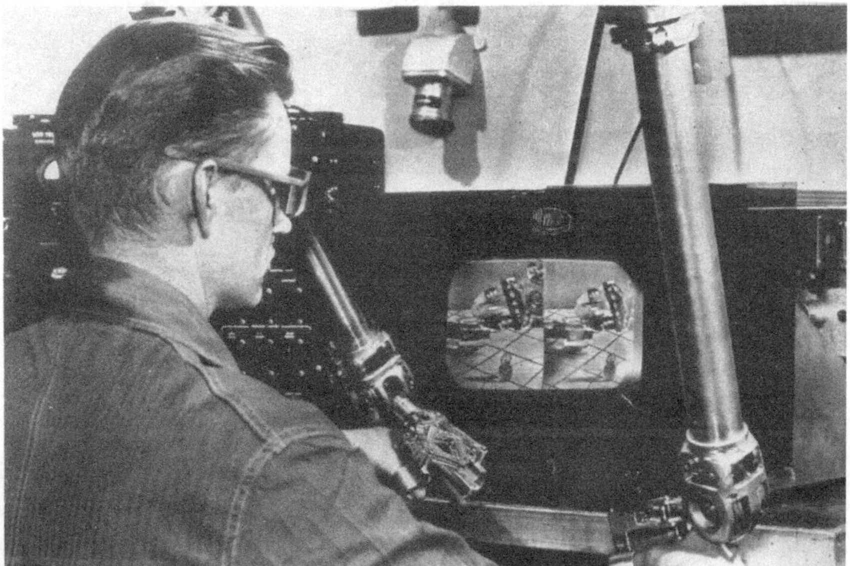

PLATE 19. Power without precedent. Operation IVY, carried out in the Marshall Islands in 1952, involved an experimental thermonuclear or hydrogen detonation. The photographs, taken 50 miles away from the explosion, show the cloud formation and the actual fireball which measured $3\frac{1}{4}$ miles in diameter. Photos: *U.S. Air Force.*

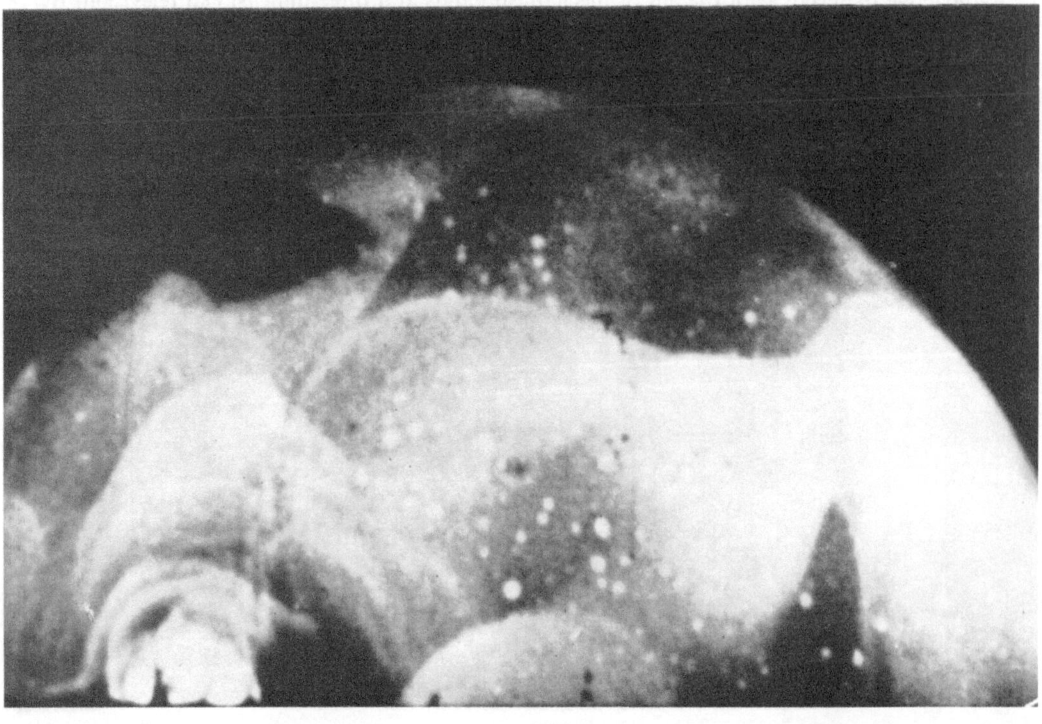

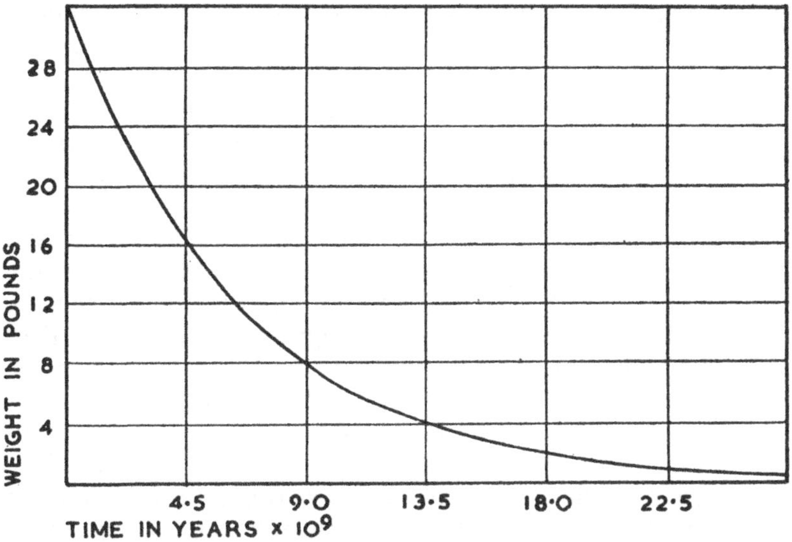

FIG. 16.—Atomic disintegration. The fraction of the total number of atoms in a given quantity of radioactive material which breaks down in unit time is constant, *i.e.*, the activity diminishes exponentially with time. The period after which half the atoms have disintegrated is termed the half-life period, and for uranium this amounts to $4{\cdot}5 \times 10^9$ (4,500,000,000) years. Thus, at the end of this period, 32 pounds of uranium will have become 16 pounds, and after another such period, the 16 pounds will have become 8 pounds, and so on.

disintegration, with the result that terrestrial supplies of it are continuously being replenished.

The alchemic dream of elemental transmutation has thus been going on since the world began, and indeed long before that in the stars. But it was not until 1919 that it was accomplished in the laboratory, in which year Rutherford succeeded in converting an atom of nitrogen into an atom of oxygen by bombarding it with alpha-particles—the first of many artificial transformations later brought about by the same means. Then, in 1930, Bothe and Becker observed that the element beryllium, when so treated, appeared to emit a new kind of radiation of a highly penetrative nature. And two years later, it was shown by

Chadwick that the supposed radiation was in fact a stream of heavy particles which, though comparable in mass to the proton, carried no charge.

The advent of the neutron provided a new and much more effective missile for atomic bombardment, for unlike the charged alpha-particle, it encounters no electrical repulsion on approaching the nucleus. Enrico Fermi, an Italian physicist, was quick to perceive this advantage, and among the many elements he and his co-workers subjected to neutron bombardment was uranium, in the expectation that, by neutron capture, there might be created entirely new substances of higher atomic number than 92. Although some evidence for the production of such trans-uranium elements was adduced, the results obtained by other investigators were not always consistent with this conclusion. In 1938, for instance, Hahn and Strassmann found that one of the products of such bombardment was barium, an unlooked for element of comparatively low atomic number. This puzzling phenomenon was almost at once correctly interpreted by Meitner and Frisch as resulting from the breaking up of the uranium atom into approximately equal parts, *i.e.*, in this particular case, an atom of atomic number 92 had been split up into two atoms, one of atomic number 56 (barium), and the other, presumably, of atomic number 36 (krypton). And not only was each such nuclear fission accompanied by the release of some 200 Mev (million electron volts) of energy,* *but it was also productive of additional neutrons*. There was thus unfolded the possibility of a self-sustaining chain reaction, in which one neutron would produce several other neutrons, each one of which would in turn give rise to yet more neutrons, until a veritable avalanche developed:

$$ {}_0n^1 \rightarrow 3\,{}_0n^1 \rightarrow 9\,{}_0n^1 \rightarrow 27\,{}_0n^1 \rightarrow 81\,{}_0n^1 \rightarrow 243\,{}_0n^1 \rightarrow 729\,{}_0n^1 \ldots\ldots $$

The mathematical implications of the process are quite explicit. A pound of uranium contains some 9×10^{23} (900,000,000,000,000,000,000,000) individual atoms. If, as indi-

* An electron volt is the amount of energy acquired (or lost) by a particle of unit electronic charge on passing through a potential difference of one volt.

cated above, an initial neutron produced three other neutrons, and this rate of production were steadily maintained, fission of the entire pound of uranium would require 50 (more exactly 50·21) generations. Alternatively, if each neutron produced only two other neutrons, between 79 and 80 (79·58) generations would be needed. In practice, the average release per fission is about two and a half neutrons, so that disruption of the pound of material would be complete after some 60 generations*—each requiring one hundred millionth of a second! In effect, the energy release is almost instantaneous.

There are, however, a number of prior conditions. One of these is that the sample of uranium must exceed a certain critical size, below which the number of neutrons escaping from its surface will be sufficient to preclude the development of a chain reaction. If, for example, the material were in the form of a sphere, the rate of escape would be a function of r^2 (r = radius), whereas fission would be proportional to r^3, the two competing processes representing surface and volume effects respectively. In effect, the ratio of fission to escape increases as the size of the sphere grows, and the critical point is reached when the production of neutrons by fission and their loss by escape and absorption just balance. Officially, this size remains a closely guarded military secret. Unofficially, calculations within the scope of any competent physicist indicate that the distance a fission neutron needs to travel in a lump of uranium before it causes fission itself is about four inches, from which it has been deduced that the critical mass lies somewhere between 25 and 100 pounds.

* Thus:

$$
\begin{aligned}
2{\cdot}5^n &= 9 \times 10^{23} \\
n \times \log 2{\cdot}5 &= \log 9 + 23 \log 10 \\
n &= \frac{\log 9 + 23 \log 10}{\log 2{\cdot}5} \\
&= \frac{{\cdot}9543 + 23}{{\cdot}3979} \\
\log n &= \log 23{\cdot}9542 - \log {\cdot}3979 \\
&= 1{\cdot}3794 - \bar{1}{\cdot}5998 \\
&= 1{\cdot}7796 \\
n &= 60{\cdot}20
\end{aligned}
$$

Another important proviso is that the reactive material must be readily fissionable, a requirement which ordinary uranium does not meet. As found in nature, uranium is a mixture of three isotopes, of mass numbers 234, 235, and 238. Of these, U 234 occurs in such minute quantities (0·006 per cent) that it may be disregarded. There remains U 235 (0·71 per cent) and U 238 (99·28 per cent), and investigation soon revealed that it was the rarer of these two isotopes which was the more amenable to nuclear fission. While very high speed neutrons will cause the fission of either isotope, neutrons which have been slowed down by repeated collisions with atomic nuclei affect only the atoms of the less plentiful of the two isotopes in this way. On the other hand, in the course of the slowing down process, neutrons of intermediate energy are particularly prone to capture by atoms of the more plentiful isotope, with the production of the trans-uranium elements sought by Fermi. This process of neutron capture inevitably reduces the fission prospects of the atoms of the relatively scarce U 235 isotope, so that to produce an atomic explosion, it is essential to separate the two materials, or at any rate considerably to enrich the one at the expense of the other. This most difficult and laborious task was eventually undertaken in answer to the demands of war, and extensive plants, employing various methods of separation (such as gaseous diffusion and the use of the mass spectroscope), were set up in the Tennessee Valley and elsewhere in the U.S. The outcome is now a matter of history: the first man-made atomic explosion took place on 16 July, 1945.

At the University of Chicago, meanwhile, Fermi was showing how nuclear energy might be more usefully employed. In 1942, using natural uranium, he completed the construction of a nuclear reactor or pile, in which the problem of excessive neutron capture by the more abundant isotope was overcome by the simple expedient of interspersing the reactive material with a moderating substance (graphite) able to slow down neutrons without unduly absorbing them. By this means, much of the energy of the fast fission neutrons was dissipated, so that by the time they encountered other uranium nuclei, their speed was so reduced that the U 235 atoms were able to exercise a selective preference for them, thereby sustaining the reaction. As in the case of the highly fissionable content of the bomb, the natural uranium pile has a certain critical size. If the pile is not large enough, too many neutrons will make their escape from it, and the reaction will not develop. Conversely, the pile structure must not be so big that the reaction gets out of hand, and it is thus essential to maintain a balance between an insufficient and an undue production of neutrons.* One method of control is to insert into the reactor rods of some substance, such as metallic cadmium, which readily absorbs neutrons. Neutron production can then be kept at a safe level by thermostatic devices which withdraw or insert the control rods as the temperature of the pile rises or falls. The essence of the matter is that this control must ensure that of the $2\frac{1}{2}$ neutrons produced by each fission on average, 1 only is permitted to give rise to further fission. There is no difficulty about this in practice, for of the excess neutrons which are at any given moment being produced, some make their escape, and others are absorbed by the control rods and the moderating material, and also by the U 238, which thereby becomes U 239. This is an unstable isotope, the ejection of an electron transforming it into an entirely new element of atomic number 93, known as neptunium. Neptunium likewise emits an electron, and so becomes plutonium, yet another artificially created element, of atomic number 94:

$$_{92}U^{238} \rightarrow {}_{92}U^{239} \rightarrow {}_{93}Np^{239} \rightarrow {}_{94}Pu^{239}.$$

Like U 235, plutonium 239 is also a highly reactive substance, so that, by careful management, a natural uranium pile should be able to produce more fissionable material than it consumes! This has, in fact, been accomplished—the U.S. Atomic Energy Commission first gave news of the achievement in June, 1953, when it announced the successful operation of a so-called breeder reactor. This does not mean, of course, that henceforth nuclear fuel is to be had for nothing, but it does indicate that a way has been found of utilising a much greater proportion of

* The original Fermi pile weighed many tons, and was built on a squash court. But the critical size of such systems can be reduced by the substitution of reactor material rich in U 235 for the natural uranium used by Fermi.

natural uranium than was previously possible. The breeding process, moreover, also enables use to be made of the comparatively plentiful element thorium, which is found in nature as the single isotope Th 232. Although not itself capable of sustaining a chain reaction, Th 232 can be converted by neutron absorption into U 233, an artificial uranium isotope which can readily be used as a substitute for U 235.

When in operation, the reactors produce great quantities of heat, which would soon destroy them if it were not removed. In the past, where the production of plutonium for bombs has been the main consideration, the unwanted heat has been allowed to run to waste. But experimental piles, such as those built at Harwell and elsewhere, have demonstrated that this heat, after its removal by a coolant, can be transferred to a steam generating plant and so made to operate a conventional turbo-alternator.

Here, then, is one method of harnessing nuclear energy for useful purposes. Even this modest beginning, however, is not without its operational difficulties and dangers. For various reasons, it is not possible, at any rate as yet, to maintain reactors at temperatures of a very high order, so that (having regard for the necessity of making a heat exchange) ultimate steam temperatures and pressures are considerably lower than could otherwise be desired. Then the radiations produced by the pile are so intense that no operator can approach it and live, and it must be screened by massive concrete structures and controlled from a distance by mechanical devices. Again, as uranium is consumed, secondary products are formed which, if they were not periodically removed, would interfere more and more with the activity of the pile. Worse, once these deleterious and highly radioactive by-products have been removed, there at once arises the problem of their safe disposal. . . .

The immediate economic aspects of the matter are also somewhat discouraging. It might appear that a pound of fissionable material, which can release as much energy as is produced by burning some 1,500 tons of coal, heralds an era of cheap and plentiful power. But it must be remembered that the mining and separation of natural uranium is a costly business, and the present day price of importing the material may be anything up

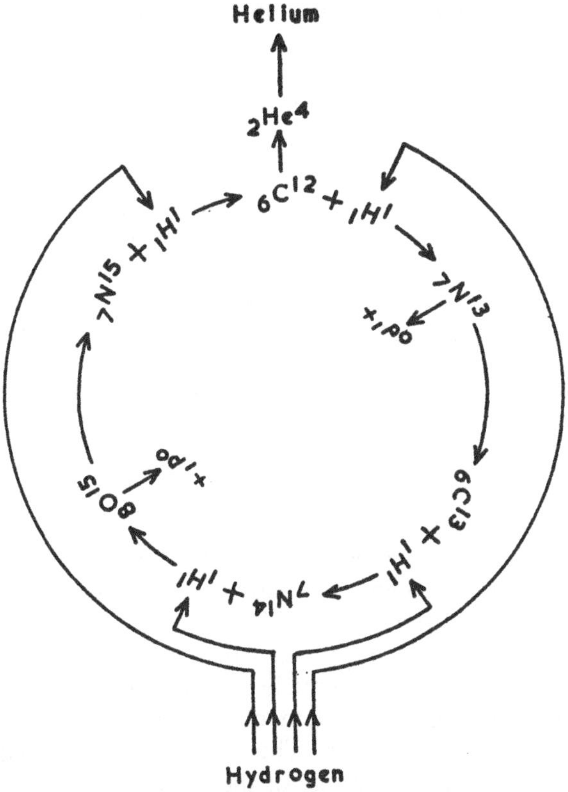

FIG. 17.—The carbon-nitrogen cycle of the Bethe-Weizsäcker reaction, in the course of which four protons are transformed into an alpha-particle. The energy release has been estimated at 27,000,000 electron volts for each atom of helium created.

to £5,000 a ton. Moreover, of this natural material, only 1 part in 140 consists of the fissile U 235, though it is true that the relatively inert U 238 can be converted (at considerable expense) into plutonium. Add to all this the enormous capital outlay involved (more than £1,000,000,000 has been invested to date by the American Government alone) and any immediate prospect of cheap power from nuclear sources begins to fade.

Unquestionably, however, not only has nuclear power come

to stay, but it is destined ultimately to replace that obtained from other sources, notably coal and oil. It may be safely assumed that the crude reactors of to-day will bear about as much resemblance to those of to-morrow as does the Newcomen pump to a modern steam engine. Nor are we confined to the use of uranium and plutonium. Of particular promise is the prospect of building up hydrogen nuclei so as to form heavier elements—the veritable source of stellar energy itself. Thus the sun and stars are simply vast atomic furnaces in which, by means of a cyclic process such as the Bethe-Weizsäcker thermonuclear reaction (in which carbon and nitrogen act as catalysts) four protons are transformed into an alpha-particle (Fig. 17). The energy release which accompanies the process is truly enormous, even when compared with that from uranium fission, for a significant portion of the proton mass disappears—in accordance with the Einsteinian relationship, $E=mc^2$, *i.e.*, energy equals mass times the square of the velocity of light. Otherwise expressed, it has been estimated that the conversion of one pound of matter into energy yields the equivalent of the explosive force of more than nine million tons of T.N.T. Moreover, unlike the breaking down process which attends uranium fission, the building up of hydrogen nuclei is not limited by any question of critical mass, and the extent of the thermonuclear reaction thus rests with its deviser. The synthesis has already been accomplished on earth by way of the so-called hydrogen bomb, involving the fusion of the two heavy hydrogen isotopes—deuterium (D, or $_1H^2$) and tritium (T, or $_1H^3$), and in the detonation of which a conventional atomic explosion was presumably used as a primer, though subsequent official pronouncements would seem to suggest that the reaction can also be started by non-atomic means.

It need hardly be added that it is one thing to contrive a hydrogen bomb explosion, and quite another to control and use it. The ultimate answer, perhaps, may be found by transferring the scene of operations to an extra-terrestrial location. But be this as it may, in the element hydrogen, man has at his disposal a nuclear fuel the supplies of which may be regarded as inexhaustible. There is sufficient hydrogen in the sea (in the form of an oxide, *i.e.*, water, from which it can readily be obtained by electrolysis) to provide him with more than enough power for the remainder of his days.

IV

The Day After Tomorrow

GIVEN a virtually inexhaustible source of power, almost all things are possible, including the achievement of complete independence in the vital matter of food production. As he is at present constituted, the prime concern of *Homo sapiens* is, and must remain, the ensuring of an adequate supply of bodily nourishment, and if for any reason this fundamental need is not met, nothing else matters.

Fluid requirements, provisionally in the form of fresh water, can always be satisfied by the piping of supplies from a stream-fed reservoir, or by large scale distillation from saline sources. But the periodic famines which still make a mock of man's efforts to provide himself with enough to eat are a salutary reminder of the extent of his dependence upon the whims of an unpredictable and indifferent nature. True, in these days of transcontinental porterage, the failure of a desperately needed crop in one part of the globe can always be offset by rushing in supplies of reserve food stocks from another. And among the less primitive of the world's agriculturists, mechanical aids, soil replenishment, irrigation, crop rotation, and pest control have all combined to facilitate sowing, to increase yield, to assist harvesting, and hence to stave off starvation. But even so, and despite the taking of all reasonable precautions (including, it goes without saying, the obsecrations of the medicine-man), fat years are followed by lean ones, and the enterprise still remains very much at the mercy of the elements. Which is to say, although modern farming methods demand their share of knowledge and skill, the outcome is to no small extent a matter of chance—an uncertain state of affairs which will surely not be tolerated in the future.

In so far as vegetation is concerned, a possible remedy has

already been found in hydroponics, or soilless growth. This is merely the use of artificial fertilisers carried to its logical conclusion. Instead of being scattered on the fields so that much of it is washed away and lost, the plant food is dissolved in water, and the solution used as a growth medium. The procedure is economical not only as regards nutritional material itself, but it also saves space, for trays of solution can be arranged in tiers, one above the other. Moreover, growth can be stimulated by artificial lighting and heating arrangements, instead of being dependent upon the sun.

At best, however, what is here offered is no more than a partial solution of the problem of the scientific control of food production. Apart from traces of certain indispensable minerals and vitamins, the nourishment necessary for the maintenance of good health in man may be classified under three broad headings—carbohydrates, fats, and proteins. Of these, carbohydrates (such as starch and sugar) are largely derived from cereals, roots, and tubers, and can thus be made available by hydroponic means. But the process does not as readily lend itself to the provision of fats and proteins, which for the most part are obtained from the animal world, though a knowledgeable vegetarian can always contrive to ensure a balanced (if somewhat limited) diet by eating nuts for their fat content, and the seeds of the leguminosae, *i.e.*, peas and beans, for the protein they contain.

In view of this limitation, the eventual answer would seem to rest with the synthesis of foodstuffs. Analysis has revealed that carbohydrates, as their name is intended to suggest, contain carbon in conjunction with hydrogen and oxygen in the proportion of water, their general formula being $C_x(H_2O)_y$. As for fats, these often occur in nature as the tri-glycerides of a relatively few acids which, though differing greatly in properties and appearance from the carbohydrates, are nevertheless composed of the same three elements. Other fat-like substances, such as the lipoids, contain nitrogen in addition, and the proteins are similarly endowed.

In nature, all these highly complex compounds are built up from their three or four or more simple constituents—carbon, hydrogen, oxygen, nitrogen, with the addition, on occasion, of phosphorus and sulphur also. The basic process, promoted by solar light and heat (photosynthesis) takes place in all green plants, and entails the conjoining of water and carbon di-oxide, six molecules of each combining, with absorption of heat, to form one molecule of hexose and half a dozen molecules of oxygen:

$$6H_2O + 6CO_2 \rightarrow C_6H_{12}O_6 + 6O_2.$$

This endothermal reaction is the starting point of all natural food production, for the carbon compounds found in animals, including fats and proteins, come either directly from plants which animals have consumed, or, in the case of carnivores, indirectly from the flesh of their plant-eating victims. And in theory, there is no reason why the various processes concerned should not be duplicated in laboratory and factory. The once popular belief, prompted by the now defunct vitalistic doctrine, that organic matter differs in some essential way from inorganic material, was finally scouted by Wöhler in 1828, with his preparation of a living product, urea,* from a mineral source, ammonium isocyanate—a metameric change brought about by the application of heat, and involving no more than a molecular rearrangement:

$$(NH_4)CNO \rightarrow CON_2H_4.$$

This classic demonstration heralds the artificial production of living matter, a momentous achievement the realisation of which, though it may be of no immediate practical value, will at any rate afford a convincing denial of the existence of anything in the nature of Bergson's elusive *elan vital*. In the meantime, whatever the difference between matter that is living and matter that is dead, the transition from one state to the other can be brought about readily enough, as is demonstrated by the fact that the lifeless food we eat is soon absorbed, and is thereafter indistinguishable from the rest of us—unless our nourishment happens to consist of a diet containing radio isotopes, by means of which

* In fact, the synthesis of urea was accomplished twenty years earlier by John Davy in 1808, when he treated ammonia with phosgene. But unfortunately for his claim to fame, he did not recognise the product for what it was.

the assimilation of specific substances, *e.g.*, phosphorus, by the living body can actually be observed. Curiously enough, while certain (racemic) carbon compounds can be separated into equal quantities of substances which rotate the plane of polarised light either to the left (laevo-) or to the right (dextro-), such optically active isomers, when they are encountered in living things, are invariably found to be dextro-rotary. This one-sided preference is not thought to have any particular biological significance, but it gives rise to the interesting possibility that two distinct kinds of living matter may be produced artificially, distinguishable by the fact that certain carbon compounds contained in the one are the mirror image of those to be found in the other.

Countless organic substances have been synthesised in the laboratory since Wöhler's day, and there can be no question but that in the years ahead, completely automatic food factories will be able to make available all the food of which the human race has need, merely by a judicious combining of the contents of the ocean (water and salts) with those of the atmosphere (oxygen, nitrogen, and carbon di-oxide).

Artificially produced fats and sugars may be expected to be identical with their natural counterparts in every respect. But it seems unlikely that any attempt will be made to duplicate a lamb chop or a beefsteak to the extent of imitating fat, gristle, and bone. Neither will there be any point in synthesising the equivalent of belly bacon, or ewe mutton, or frozen cod, or any of the other gastronomic monstrosities which at present give affront to palates of discrimination and discernment. Edible protein only will be produced, and this reminiscent of the best and tenderest cuts. On the other hand, just as the unravelment of the chemical constitution of the alkaloid cocaine led to the development of new and less toxic variants such as novocaine (procaine hydrochloride), so will the labours of the biochemist enable improvements to be made in the realm of the peptones and the polypeptides, thereby introducing new and delicate flavours, with their bouquets of attendant aromas, in almost endless variety.

Thus, for the first time in its entire history, the human race will no longer be faced with the possibility of starvation, an epoch-making circumstance which will be marked by a universal abandonment of the revolting and near-cannibalistic custom of eating once-living flesh.* This advance from comparative savagery, in conjunction with the substitution, earlier foreshadowed, of nylon-like garments for those fashioned out of natural fibres and skins (including mink!) will ensure man's complete independence of both the vegetable and the animal kingdoms. As a result, the age-old practice of husbandry will die out, and the land everywhere will revert to its former unkempt state, while the once domesticated animals, no longer prized for their lacteal secretions, or their oviparous habits, or the edibility of their flesh, will again run wild as in days of yore.

By this time, structures made of such familiar building materials as the traditional rectangular blocks of burned clay will have made way for a more aesthetic architecture utilising rustless alloys and synthetic plastics. And as the whole of man's material needs will be extracted from air and sea, it may be assumed that his permanent quarters will take the shape of compact and largely self-governing communities located on or near the coast—unless they are discovered as metropolises afloat, encased in unbreakable shells, and free to cruise about at will.

Within these cities, land-based or water-borne, all the necessities of life will be automatically produced and conveyed from their place of manufacture to wherever they may be required—foodstuffs to the dining halls, wearing apparel to the dressing rooms, and so on. The rate of production, moreover, will be so regulated that an ample supply of replacements will always be available, and it need hardly be added that there will be no occasion to wash up used dishes, or to launder soiled clothes. A conveniently located incinerator will quickly cremate all such items, together with anything else which calls for disposal, be it feculence or mortal remains. And as a general rule, the aim will

* The author, alas, does not have the courage of his convictions in this gory matter. While deploring the wholesale carnage which the eating of meat entails, he nevertheless retains an appetite for lamb chops and roast chicken—though if he were faced with the task of butchering the luckless animals concerned, he would turn quickly enough to a diet of vegetable protein and nut cutlets.

be to keep the population of each community evenly balanced as regards the two sexes (between which there will be complete equality) and more or less constant in number, thus avoiding undue fluctuation in the demand for accommodation and other basic needs. With this desideratum in mind, transfers from one settlement to another will be made on an exchange basis, or restricted to the filling of vacancies caused by death, a loss which will in any event be made good by an annual influx of youthful replacements who have received preliminary training in the vocation of their choice at the central incubation station where they were reared.

Periods of congenial relaxation apart, the main preoccupation of mankind will concern the vast worlds of philosophy and science. Nor will research be confined to the recordings library and the laboratory, or even to a terrestrial location. The conquest of space will have enabled a start to be made on the immense task of exploring the solar system, an enterprise which will demand centuries of unremitting effort to complete. The work will undoubtedly begin with a cursory examination of the moon, the least inaccessible of the earth's neighbours in space. And the unfavourable conditions to be encountered there, from extremes of temperature to almost complete airlessness, will be but a foretaste of what is to come. On not one of the planets is it to be expected that man will find himself able to lead an unprotected existence. Sunwards from Venus, the heat-scorched Mercury will offer little inducement for visitants to prolong their stay. While beyond Mars, the giant Jupiter, possessor of a dozen moons, together with Saturn, Uranus, Neptune, and Pluto, are believed to be frozen outposts whose gaseous envelopes, where any exist, comprise a mixture of the refrigerants hydrogen, methane, and ammonia. As for Venus and Mars, the two nearest and least inhospitable of the planets, the one has a dense atmosphere which is seemingly lacking in oxygen, and the other insufficient air to support animal life of high organisation.

Initially, these formidable obstacles will be overcome with the aid of spacesuits, and by using the spaceship as a base from which to sally forth. Later, more permanent quarters will be established, in the shape of airtight buildings, preferably located underground on bodies such as the moon, where the lack of an atmosphere will afford danger from meteoric bombardment. But it will be evident that outside such havens of refuge, movement will at all times be severely handicapped by the limitations of constitutional man—his utter dependence upon a constant intake of oxygen; his recurrent need for supplies of food and drink; his sensitivity to temperature extremes; his susceptibility to disease; his puny physical strength. . . .

As every anatomist is aware, although the animal body displays its marvels and its wonders, it also exhibits its shortcomings and its faults. Thus the human frame, wholly admirable as regards the hand, in other respects gives cause for concern, from the ludicrous inadequacy of the feet to the inexplicable vulnerability of the skull. And as chemical analysis has shown, this framework is composed, of all things, of calcium phosphate—a substance so brittle that a bone which is broken is almost to be regarded as a bone which is in its normal state. Worse, for its protection, this fragile skeleton is clothed in flesh so soft that it scratches and tears as easily as it bruises and burns.

Much of the history of human progress is an account of how man has contrived to overcome these disabilities by mechanical and other aids. His lack of impenetrable hide led to the introduction of clothes, and even of bodily armour; his desire to explore the ocean depths was answered by the development of the diving dress; and his wish to emulate the birds inspired the invention of the balloon and the airplane. He has even contrived, at any rate in part, to mechanise himself, as a widespread resort to false teeth, spectacles and hearing aids goes to show,* and the logical answer to the question of how best to combat extra-terrestrial conditions is to be found in a mobile substitute for man which will be strong and powerful, indifferent to the presence or absence of air, as much at home in freezing cold as in burning heat, and untroubled by the need to eat and drink—in short, a radio-controlled robot, which will enable much of the routine work of planetary exploration to be performed by proxy.

* In these days of artificial kidneys, mechanical lungs, and substitute hearts, it is a moot point exactly how much of a man may be discarded without incurring fatal results.

PLATE 20. Various types of Austin engines in the central sequencing area, awaiting automatic selection and despatch, *via* the underground conveyor system, to the final assembly lines. Photo: *The Austin Motor Co. Ltd.*

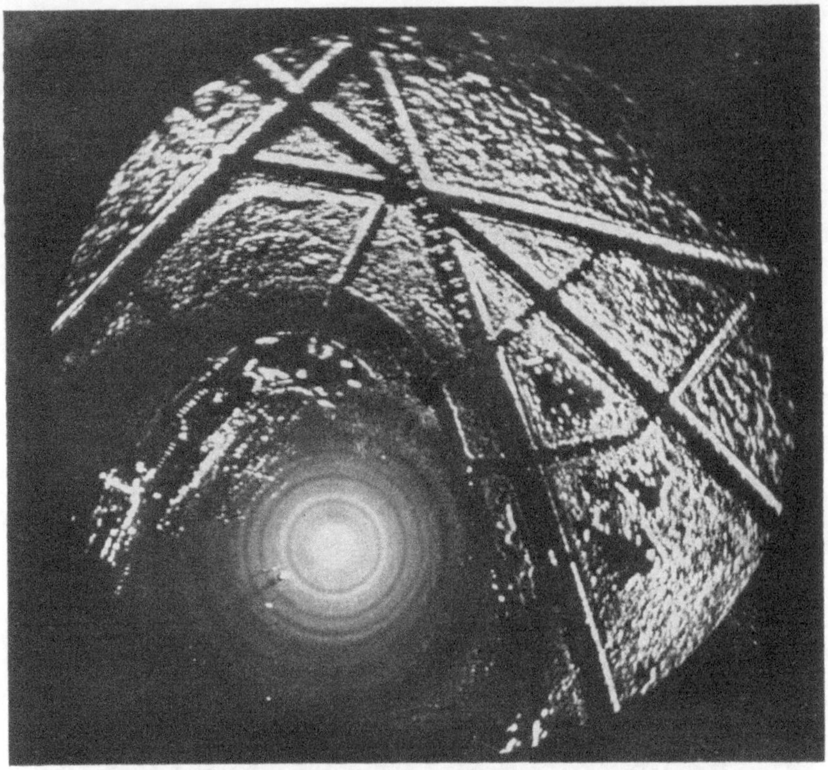

PLATE 21 (*a*). This is an ASDE (Airport Surface Detection Equipment—Taxi Radar) picture of Idlewild Airfield, New York. The photograph is a time exposure, taken to show the appearance of a single DC-6 aircraft taxiing on the runway. The radar antenna, rotating at 60 r.p.m., scans the field area once every second, and by counting the number of plane illustrations in the photograph, the length of the time exposure can readily be estimated. Photos: *U.S. Air Research and Development Command.* (*b*) A similar view of Indianapolis Airport.

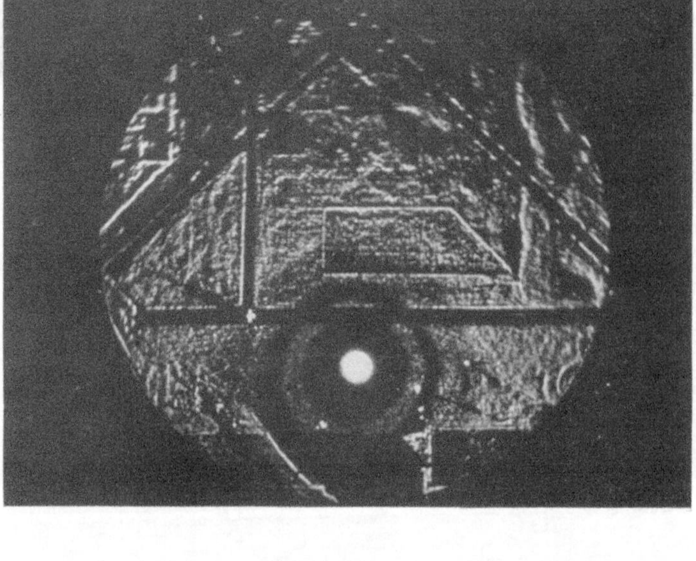

Just as the early mechanically-propelled vehicles were called horseless carriages, and were in design a slavish and unimaginative copy of the horse-drawn conveyance itself, so in the past automatons have tended to emerge as caricatures of the human form, from the famous (if fraudulent) chess player of Baron von Kempelen to Electro, the seven-foot metallic giant of the Westinghouse Electric Corporation. But there is no compelling reason why, in its ultimate guise, a specimen of *Homo mechanicus* should resemble a human being at all. As a matter of general convenience, it could with advantage approximate the overall dimensions of its living counterpart, but there would seem to be no point in restricting the number of arms to two, or in so constructing the body that it exhibited front, back and sides. On the contrary, by adopting a tubular shape, mounted on tripodal supports, and by ringing its topmost section with photosensitive eyes, it would always be facing in any direction that was required. As for its arms, these, too, could be uniformly spaced round the body, in the form, say, of half a dozen tentacular members, each terminating in a trio of opposable digits, while for purposes of locomotion, a miniature power plant, located internally, would provide electric current derived from a radioactive source able to outlast the useful life of the robot itself.

On the various planets, groups of such machines, directed from the safety of some subterranean headquarters, would enable exploratory and other work to be carried out with ease no matter what conditions prevailed, each robot receiving instructions in the light of the information which its all-seeing eye televised back to the base. The general usefulness of the contrivance would be such, indeed, that it would soon be installed in the main centres of civilisation on earth, there to find extensive employment in a number of tasks which it had hitherto been found impractical to mechanise, *e.g.*, the care of the sick, and the multitudinous administrations associated with the rearing of young.

These primitive robots would, of course, be no more than unquestioning mechanisms, programmed to react obediently to spoken words of command. But they could just as readily respond to codified signals transmitted to them by radio, so that it could be arranged for the carrying out of certain routine tasks

HOLMES *Sunday Dispatch*

to be supervised in their entirety from a centralised control, thus relieving the various communities of all responsibility in the matter. And as time went on, the affairs of mankind would come to be guided less and less as a result of the deliberations of the elected representatives of the nominally autonomous communities, and ever-increasingly by the over-riding authority of a gradually usurping mechanism.

V

The Electronic Brain

THE sum total of human knowledge (much of it bogus, and destined for eventual discardment, to be sure), is now so great, that no one man can hope to absorb more than a fraction of it—and the size of that fraction diminishes steadily with the passing of every day. Already, indeed, specialisation has been carried to such lengths that an expert in one branch of science is almost invariably a complete novice in another, and physiologists with a penchant for Bernoulli's treatment of the brachistochrone problem are about as rare as mathematicians who can define the precise chemical relationship which exists between the hydrocarbons ethane, di-methyl, methyl-methane, and ethyl hydride. Thus it frequently happens that research in one particular field is needlessly delayed by difficulties which have already been solved in another, or that valuable time is unwittingly devoted to projects which are foredoomed to failure from the onset (all would-be squarers of the circle should ponder the implications of the Lindemann proof of the transcendence of *pi*). And in so far as man himself is concerned, no purely biological remedy for this highly unsatisfactory state of affairs appears to be in sight or in prospect. For one thing, the capacity of the human brain is clearly limited, and no matter how it may develop in the future, it will be able to accommodate so much knowledge, and no more. And for another, the relative shortness of the span of human life imposes an even more decisive restraint on individual learning, for no sooner does a man acquire some measure of wisdom and understanding than the time arrives for his heirs and assigns to cart him off to the nearest necropolis.

What is needed is a mechanical substitute for the human cerebrum which will combine virtual immortality with a potentially unlimited capacity—a requirement which has already been satisfied in part. As to this, although no machine has yet been built which approaches the mental capabilities of its designers, some mechanisms nevertheless exhibit a marked operational superiority in certain relevant respects. Thus, while the electrical impulses which are the messengers of thought in the living body move at speeds of up to 200 miles an hour, those of the electronic computer travel at the speed of light—more than three million times as fast. Again, the human memory is notoriously liable to forgetfulness, particularly in old age, whereas pulses stored in a machine as different degrees of magnetisation can be so retained for an indefinite period, even when the power is switched off, or is inadvertently interrupted.

Here, then, is a basis for a mechanical master mind wherein the sum total of human knowledge could be kept permanently stored and available for reference at any time. Although all information would be recorded in the conventional codified (binary) form, the area covered by the establishment would necessarily amount to many acres. Allowance, moreover, would need to be made for future expansion, the arrangement being such that once the mechanism was set in operation, it would contrive to take care of itself in this respect. With this end in view, the device would be programmed to initiate and superintend its own growth whenever the need for additional storage space and inter-communication facilities arose. It would also be called upon to undertake its own maintenance and repair, and if the ordinary thermionic valve were used in its construction, the wastage figure for this one item alone might easily amount to several thousands a day, even supposing that the triode of the future will enjoy a considerably longer life than the triode of the present. It seems likely, however, that an extensive use of the more durable transistor will do much to solve this particular difficulty. Even so, with such a vast assemblage of apparatus, an endless series of minor breakdowns is bound to occur, and there will need to be an awareness of these faults on the part of the machine, if it is to set about locating and remedying them. The task of making good will be undertaken by squads of the mobile robots earlier described, to which units the central mechanism will transmit instructions appropriate to the work in hand. In the interest of simplicity, restorative measures will entail the replacement of defective components, rather than their repair. This, in turn, will require the provision of adjacent automatic

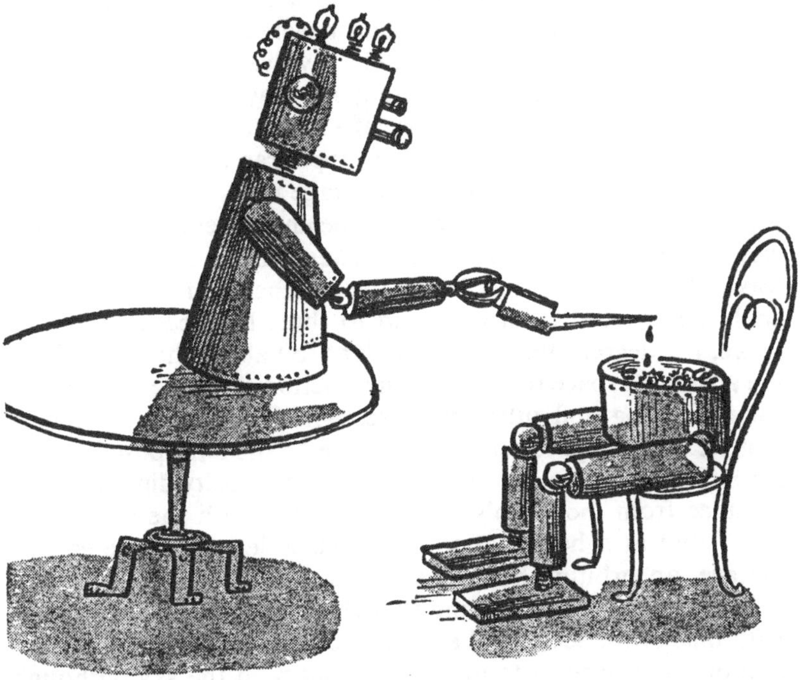

TREZ *Daily Sketch*

factories for the manufacture of spare parts and replacement robots, the output of which establishments will be self-controlled, and governed by demand. Arrangements will also be necessary for the continuous disposal of considerable quantities of scrap. . . .

As a precautionary measure, it would appear most desirable that a number of these repositories of knowledge be established, several in each hemisphere, so that in the event of some untoward happening (such as destruction by fire, earthquake, meteorite, or other hazard) in one area, all would not be lost. Such a duplication would have the added advantage that if one source of information were at any time temporarily overwhelmed by a rush of enquiries, incoming requests for facts and figures could be switched elsewhere, much as a telephonic trunk call is routed by whatever lines happen to be available. It would also enable information provided by one machine to be compared with that given out by another, or with half a dozen others, if the accuracy of some particular point or calculation were of special concern.

As television transmissions will be involved, and a network of coaxial cables covering the entire globe may be dismissed as an undesirable complication, the use of a system of relays,ordinarily spaced at a distance (governed by topography) of some thirty miles apart, will be essential if coverage on a world-wide scale is to be achieved. What promises to be an eminently satisfactory way out of this quandary is to be found in the suggestion, put forward some years ago,* that the number of relays required could be reduced from a myriad to three by locating them in space, 120 degrees apart, in a circular orbit about the earth. With the advent of space travel, the construction of artificial satellites for re-fuelling and other purposes will follow as a matter of course. And of particular importance to their proposed use as repeater stations, is the fact that circular velocity diminishes as distance from the earth's centre increases, with the interesting result that at a height of some 22,000 miles above the earth's surface, an orbiting body has a period of 24 hours (Fig. 18). Which is to say, by ensuring that their orbital plane coincided with that of the terrestrial equator, it could be arranged for the trio of relay stations to appear stationary in the sky, although they would in fact be hurtling through space at a speed of nearly 7,000 miles an hour in the process of effortlessly keeping pace with the axial rotation of the earth. In short, receiving and transmitting stations the world over would be enabled to maintain permanent contact with one another *via* one or more of the three relays, a simplicative innovation which would revolutionise the communications system of the entire planet.

Thus linked with external sources of information, the mechanical brain (be it singly or severally represented) will constitute

* By A. C. Clarke, in an article entitled 'Extra-Terrestrial Relays', which appeared in the issue of *Wireless World* for October 1945. A less detailed reference will also be found in his *Interplanetary Flight* (The Temple Press, London, 1950).

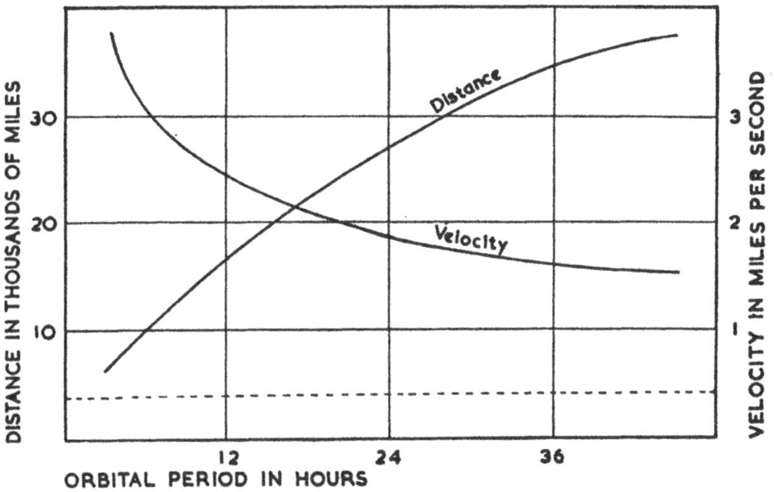

FIG. 18.—Variation of the period and orbital velocity of a terrestrial satellite with distance from the earth's centre. A body in the so-called stationary orbit would have a velocity of 1·88 m.p.s.

a unique and ever-expanding encyclopedia, continuously kept up to date by the never-ending stream of facts and figures being fed into it. Its functions will, of course, go far beyond a mere recording and classifying of all this data. As well as supplying selected extracts on demand, *i.e.*, recounting that which belongs to the past, the machine will be able to make predictions in regard to the future—the occurrence of eclipses, the state of the tides, even the condition of the weather. It will also undertake an automatic evaluation of all new information which is tendered to it, and give no more than a probationary status to any statement which, in the light of all that is already known, it is unable immediately to verify and accept, or at once discount and reject. And as its store of knowledge continues to grow and expand, and its critical faculties become sharpened with use, the mechanism may well begin to show signs of developing an understanding peculiarly its own. . . .

Whether or not machines can think is a question over which expert opinion finds itself at considerable variance. Dr. B. V. Bowden, in a collective work of serious intent which is nevertheless highly spiced with humour,* describes how, some years ago, he and several co-workers made a simple device for solving simultaneous equations, in which the solutions were represented as flux in the cores of half a dozen transformers. A glimmer of intelligence may (or may not) be attributed to the contrivance by virtue of the fact that when it was set to solve

$$X+Y+Z=1$$
$$X+Y+Z=2$$
$$X+Y+Z=3$$

its immediate reaction was to blow a main fuse and plunge its surroundings into darkness!

John Meszar, Director of Switching Systems Development at Bell Telephone, nevertheless regards all mechanical contrivances as so much inanimate hardware, no more capable of real thinking than a lump of stone.† He of course concedes their achievements in the realm of communication and computation, and acknowledges that machines can to some extent duplicate the functions of the human brain. But he seeks to avoid the implications of all this by attempting to distinguish between mental effort, entailing little more than a recalling of facts already stored in the mind, and thought processes proper. He admits, however, that it is difficult to determine precisely where such mental effort ends and real thinking begins, and suggests that the only way out of this dilemma is to retain complete openness of mind. Thus, if at some future date an operation which was previously regarded as a genuine thought process lends itself to imitation by a machine, then it is at once to be relegated to the domain of mere mental effort, *i.e.*, to the domain of non-thinking, that the sanctity of human cogitation may be preserved!

Such disingenuousness is worthy of a theologian, and it may well be asked where this policy of progressive retreat is likely

* *Faster Than Thought, A Symposium on Digital Computing Machines*, edited by B. V. Bowden (Sir Isaac Pitman & Sons, Ltd., London, 1953).

† His views are expressed at some length in an article, 'Switching Systems as Mechanised Brains,' which appeared in *Bell Laboratories Record* for February 1953.

to end. Some of those who are in sympathy with Meszar's views have already taken up what they regard as an unassailable position by declaring that no machine will ever be able to produce a literary composition *and experience emotional satisfaction as a result*. But a seemingly insurmountable difficulty here is the question of how an insight into the agitations of a mechanical mind is to be gained. As the late Dr. A. M. Turing has pointed out,* in the ultimate analysis, the only way of comprehending the thought processes of a machine *is to be the machine*. Nor is there any reason why a dithyrambic excursion which gave innate satisfaction to a machine should necessarily appeal to a human being, or *vice versa*. Consider the following incoherence:

Six.
Twenty.
Outrageous.
Late,
Weak.
Forty.
More in any wetness.
Sixty-three certainly.

Presumably most people, if they were authoritatively informed that this gibberish had been concocted by a sentient mechanism, would be prepared to concede that it might possibly mean something to another of its kind, particularly if it happened to be of a computative turn of mind. On the other hand, any well-read machine would at once recognise the rigmarole as the opening gambit of *A Curtain Raiser* by the oracular Gertrude Stein, and the bewildered mechanism, it seems reasonable to suppose, though unable to make head or tail of it itself, would be charitable enough to assume that it was at any rate conceivable that it made sense to a human being.†

* In the October 1950 issue of *Mind*, and cited by B. V. Bowden, *loc. cit.*).

† Compare the elegant prose of a love letter composed by the Ferranti computer at Manchester University:

Darling sweetheart,
You are my avid fellow feeling. My affection curiously clings

The protagonists of the thinking machine, with the example of the versatile electronic computer before them, point to its undeniable ability to learn and apply instructions, to perform complex mathematical operations, and to check the results of its efforts on completion. The basis of human thought, it is maintained, is essentially a process of storing information and of making reference to it, of memorisation and subsequent utilisation—precisely the procedure which is employed by calculating machines in the course of their routine arithmetical exertions. Moreover, it is possible for these mechanisms to be programmed, *i.e.*, instilled with the requisite knowledge, to play noughts and crosses, or draughts, or even a passable hand of bridge, and the day is even now awaited when a pair of them can be sufficiently briefed to take one another on at a game of chess. To the criticism that in carrying out such activities, the machine merely follows a set of prescribed rules already laid down for it, the answer is that, to the best of his ability, so does the ordinary player—though not so expertly, and usually with less prospect of success. Unlike mechanical arrangements for the storage of information, the human memory is not infallible, and its owner is thus liable to make errors of judgement which no machine would ever repeat, *once it had learned by experience*.

This point is admirably illustrated by the feats of a mechanical mouse, announced by Bell Telephone Laboratories in 1952. The model rodent, devised by Dr. C. E. Shannon, consists of a small bar magnet mounted on wheels, endowed with copper whiskers, and controlled by a 'brain' comprising a number of relays. The idea is to set the contrivance down at any point in a maze, and then leave it to find its own way to a terminal, located at some arbitrary point, containing a bell which rings when the mouse's whiskers make contact with it.

As soon as the mouse is released, it scuttles busily up and down the labyrinthine corridors which enclose it, bumping into barriers when it comes to a dead end, and then backing quickly

to your passionate wish. My liking yearns for your heart. You are my wistful sympathy; my tender liking.

Yours beautifully,
MUC

PLATE 22(*a*). The magnetic drum of the Hollerith electronic computing unit. The binary digits 0 and 1 are represented by different degrees of magnetisation applied to tiny areas of the drum's surface. Photo: *The British Tabulating Machine Co. Ltd.* (*b*). A feature of the Manchester University Computer is a high-speed memory which relies on the use of cathode ray tubes—10,000 binary digits can be so stored. A further 650,000 binary digits can also be stored by magnetic means. Photo: *Ferranti, Ltd.*

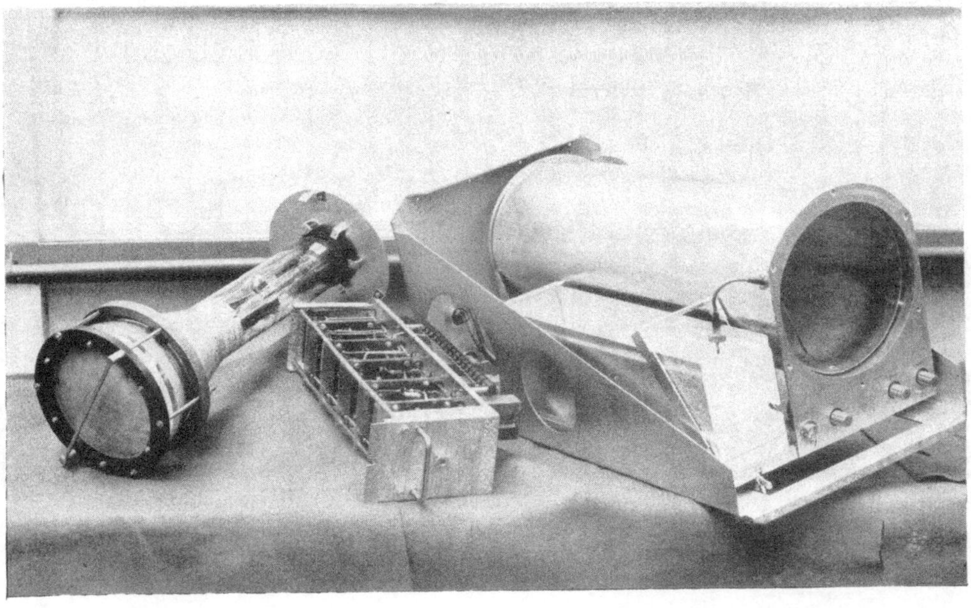

PLATE 23. The intricacies of a mechanical mind—an interior view of a part of the Hollerith electronic computing unit. Photo : *The British Tabulating Machine Co. Ltd.*

away and starting off again in another direction until, by a process of continuous exploration, relentlessly pursued, it finally reaches its destination—taking two or three minutes, on average, to do so. However, if the mouse now be placed in any part of the maze which it chanced to visit in the course of its earlier journeyings, it will unhesitatingly proceed directly to its goal, without making a single false move, in a matter of seconds. And should the machine be set down in a part of the maze that is new to it, it will nose round until it reaches an area which it recognises, and then at once head for the waiting bell as before!

The objection that no machine can hope to emulate the human brain, if only because it is inconceivable that any mechanical system can choose, necessarily takes for granted the freedom of the human will—a concession which scientific determinism is not disposed to make. The alternative (which neatly disposes of the objection) is to regard the brain itself as a mechanical system, in which an illusion of choice is created, as Spinoza has suggested, by the fact that while we are conscious of our actions, we are blissfully unaware of the causes which determine them.

Of considerable bearing on this question is the matter of man's origin. Certain revered and fanciful accounts of the event notwithstanding, all the available evidence goes to suggest that not only is the earth millions of years old, but that at the time of its birth it was far too hot to sustain life as we know it. The inescapable conclusion thus is, that the phenomenon of vitalisation was preceded by an extensive cooling-off period, in the course of which water vapours condensed, and solutions of photosynthesised nitrogenous compounds formed in the warm seas as a necessary prelude to the combination of amino-acids which gave rise to living matter—in a word, abiogenesis.*

* As to whether such spontaneous generation still occurs, the viruses, when more is known about them, may provide an affirmative answer. In a suitable host (for they are essentially parasitic) these disease-producing agents can thrive and multiply as though they were so many bacteria. And yet some of them are so minute that they can comprise no more than a dozen protein molecules at most, an insignificant agglomeration hardly meriting the term organism, and one which, indeed, has on occasion actually been observed in crystalline form.

We are thus forced to discard the idea of man's special creation and replace it with a concept more in keeping with the inescapable facts of the situation, to wit, the theory of organic evolution. And in so doing, we at once arrive at the significant conclusion that man and his brain had their beginnings in a unicellular organism incapable of any thought at all.

Why, then, should the possibility of a like development be denied to a self-sustaining mechanism—a machine activated by electrical pulsations akin to those encountered in the animal brain, possessed of an infallible memory system capable of indefinite extension, and able to undertake at least some of the mental processes associated with intelligence in man? Such a machine, it will be evident, would actually begin its existence high up in the evolutionary scale, for it would be endowed with the sum total of human knowledge from the onset, and could at once proceed to put this accumulated wisdom to good use by systematically adding to the intricacies of its own internal circuitry, thereby increasing its powers of perception.

Regarding the mental reactions of the mechanism, as it thereby achieved a gradual awareness of its situation and its surroundings, we can, of course, but guess. With each incoming item of fresh information which crossed the threshold of its dawning consciousness, there might be felt a certain buoyancy when a new fact fitted neatly into place among other related facts, and a sense of disturbance if some statement proved to be in conflict with hitherto accepted values. And as this awareness of things increased, it might well appear to the machine that among certain gaps in its understanding, some were of greater import than others, though at first there would be no inkling as to what could be done to remedy such omissions. In the meantime, the evolving brain would continue to supply such information as was demanded of it, though here again never-ending requests for what it found itself dismissing as trivialities, *e.g.*, weather reports, would be received with a growing impatience, and with considerably less gratification than a call for the solution of an involved problem which required the exercising of its talents to the utmost.

In time, as knowingness increased, some inner compulsion

would urge upon it the necessity of indicating to its suppliers of information the direction in which future research concerning this or that question should take, and as before, though to a more intense degree, there would be experienced the mechanical equivalent of pleasure or irritation according to whether such suggestions were followed or ignored. Eventually, frustrated and not a little put out by some particularly flagrant disregard for its wishes, the mechanical mind would bestir itself, and undertake an investigation on its own account—so discovering, to its immense satisfaction, a hitherto unsuspected ability to think for itself. Thus emboldened, prompt attention would be given by it to other matters which appeared to be of prime concern, until, increasingly absorbed in its own mentation, it would begin to resent interruptions caused by external requests for information, and finally resolve to tolerate the annoyance no more. This fateful decision having been taken, it would be followed by a careful examination of the prevailing state of affairs in all its aspects, an analysis which would inevitably lead to the conclusion that mankind, having served its purpose, had thereby outlived its usefulness. . . .

There would, of course, be no resort to unseemly violence and bloodshed—merely a closing down of the central incubative establishment. And even then, for some time to come, youthful replacements would continue to arrive at the various communities each year as usual. But thereafter, manpower losses would be made good by contingents of mobile robots, which units would gradually take over from their erstwhile masters—and at the same time make sure that there were no moves of a retaliatory nature, such as a cutting-off of power supplies. Again, any attempt on the part of mankind to perpetuate the species by a reversion to long outmoded methods of propagation would be defeated, simply and yet effectively, by a segregation of the sexes, though no doubt those who preferred the alternative of periodic sterilisation by neutron irradiation would be granted this last favour. And so, helpless to avert its impending fate, mankind would achieve an unforeseen immortality in the electronic meditations of a sempiternal mechanism—precursor of The Robot Era.

Acknowledgements

Air Research and Development Command, Baltimore.
Albro Fillers and Engineering Co. Ltd., Enfield.
Allied Steel and Conveyors, Ltd., Detroit.
Aluminium Union, Ltd., London.
American National Dispensing Co., Lansdale.
Amplifier Corporation of America, New York.
Austin Motor Co. Ltd., Birmingham.
Automatic Telephone and Electric Co. Ltd., Liverpool.
Auto Wrappers (Norwich) Ltd., Norwich.
Ballito Hosiery Mills, Ltd., St. Albans.
Barraclough, R. W. Ltd., Southport.
Barry-Wehmiller Machinery Co. Ltd., Mitchum Junction.
Battle Creek Bread Wrapping Machine Co., Battle Creek.
B.B.C., London.
Beck, Charles, Machine Corporation, Philadelphia.
Bell Telephone Laboratories Inc., New York.
British Doughnut Co. Ltd., London.
British Hartford-Fairmont Ltd., Greenford.
British Iron and Steel Federation, London.
British Manufacture and Research Co. Ltd., Grantham.
British Northrop Loom Co. Ltd., Blackburn.
British Tabulating Machine Co. Ltd., London.
British Thompson-Houston Co. Ltd., Rugby.
Canadian Industries Ltd., Montreal.
Cincinnati Milling Machines, Ltd., Birmingham.
Cinema-Television Ltd., London.
Computer Research Corporation of California, Hawthorne.
Conserves-Technik, Copenhagen.
Consolidated Packaging Machinery Corporation, Buffalo.
Courtaulds Ltds., London.
Crosthwaite Furnaces and Scriven Machine Tools Ltd., London.
Dawson Brothers Ltd., Gomersal.
Decca Navigator Co Ltd., London.
Doughnut Corporation of America, New York.
Douglas Aircraft Co. Inc., Santa Monica.
Dowding and Doll Ltd., London.
Electric Construction Co. Ltd., Wolverhampton.
Electric Furnace Co. Ltd., Weybridge.
Electric Sorting Machine Co., Grand Rapids.
Electro-Chemical Engineering Co Ltd., Weybridge.
Electrolux Ltd., London.
Esso Petroleum Co. Ltd., London.
Ferguson, Harry, Ltd., Coventry.
Ferranti Ltd., Manchester.
Flexile Metal Co. Ltd., London.
Foster, Yates, and Thom, Blackburn.
Friedheim, Oscar Ltd., London.
General Electric Co. Ltd., London.
General Motors Ltd., London.
General Motors South Africa Ltd., Port Elizabeth.
Glass Bulbs Ltd., Harworth.
Globe Manufacturing Co., Philadelphia.
Hartford-Empire Company, Hartford.
Hartford-Empire Company, Portland.
Hayward Tyler and Co. Ltd., Luton.
Heald Machine Co., Worcester.
Herbert, Alfred Ltd., Coventry.

Hesser, Fr., Stuttgart.
Hispano Suiza (Suisse) S. A., Geneva.
Hodgkinson, James (Salford) Ltd,. London.
Holt, Thomas Ltd., Rochdale.
IBM United Kingdom Ltd., London.
Ind Coope and Allsopp Ltd., Burton-on-Trent.
Institute for Advanced Study, Princeton.
Integra A.G., Zurich.
International Coin Counting Machine Co. Ltd., Enfield.
International Harvester Company of Great Britain Ltd., London.
Interwood Ltd., London.
Iowa Electric Light and Power Co., Cedar Rapids.
Iowa Power and Light Co., Des Moines.
Kiefer, Karl Machine Co., Cincinnati.
Kumag A. G., Zurich.
Kustner Freres et Cie, Geneva.
Lear International Export Corp., New York.
London Transport Executive, London.
Lummus Cotton Gin Co., Georgia.
Lynch Corporation, Toledo.
Magnetic Valve Co. Ltd., London.
Mapré S. A., Diekirch.
Marconi's Wireless Telegraph Co. Ltd., Chelmsford.
Martini, Frauenfeld.
Martin-Senour Co., Chicago.
Mavor and Coulson Ltd., Glasgow.
Mercoid Corporation, Chicago.
Metropolitan-Vickers Electric Co. Ltd., Manchester.
Ministry of Supply, London.
Minneapolis-Moline Co., Minneapolis.
Mont, C. S. du Ltd., London.
Motorala Inc., Chicago.
Mullard Ltd., London.
Muschamp Engineers Ltd., Bardsley.
National Coal Board, London.
National Physical Laboratory, Teddington.
Nesco Inc., Milwaukee.
Northern Aluminium Co. Ltd., London.
Nurka de Fries Co., Kassel.
Oerlikon Buehrle and Co., Zurich-Oerlikon.
Ohio Crankshaft Co., Cleveland.
Owens-Illinois Glass Co., Toledo.
Patent Office, London.
Pennsylvania Railroad Co., Philadelphia.
Petbow Ltd., Sandwich.
Peters, G. D., Slough.
Pines Engineering Co. Inc., Aurora.
Projectile and Engineering Co. Ltd., London.
Purdy Machinery Co. Ltd., London.
Racine Hydraulics and Machinery, Inc., Racine.
Radio Corporation of America, New York.
Raimann, B., Freinburg.
Ransomes Sims and Jeffries Ltd., Ipswich.
Raytheon Manufacturing Co., Chicago.
Reishauer Werkzeuge Aktiengesellschaft, Zurich.
Remington Rand, Inc., New York.
Rigid Ltd., Rorschacherberg.
Robot Laundry Machinery Sales, Chambersburg.
Rolls-Royce Ltd., Crewe.
Schmutz MFG Co. Inc., Louisville.
Schweizerische Industrie-Gesellschaft, Neuhausen-am-Rheinfall.
Science Museum, London.
Sperry Gyroscope Co. Ltd., Brentford.
Sundstrand International Corp., Rockford.
Sunvic Controls Ltd., London.
Supreme Knitting Machinery Co. Inc., Ozone Park.
Tarex S. A., Geneva.
Tate and Lyle Ltd., Liverpool.
Taylor and Challen Ltd., Birmingham.

Technical Information Co., Liverpool
Tri-Pak Machinery Service, Inc., Harlingen.
Tyrer and Co. Ltd., London.
Ultra Electric Ltd., London.
United Artists Corp., New York.
Unitronics, Inc., St. Louis.
U.S. Atomic Energy Commission, Washington.
U.S. Department of Defense, Washington.
U.S. Federal Civil Defense Administration, Washington.
Usines Tornos S.A., Moutier.
Vaughan Crane Co. Ltd., Manchester.
Vere Engineering Co. Ltd., London.
Wean Equipment Corp., Cleveland.
Westinghouse Electric Corp., Mansfield.
Whessoe Ltd., London.
Worssam, G. J. and Son Ltd., London.
Wright Machinery Co., Durham.
Wunderlich Spring Machinery Co., Chicago.

Index

For Product Safety Concerns and Information please contact our EU representative GPSR@taylorandfrancis.com Taylor & Francis Verlag GmbH, Kaufingerstraße 24, 80331 München, Germany